MEETINGS WITH MOTHS

KATTY BAIRD

Meetings with Moths

Discovering their Mystery and Extraordinary Lives

4th ESTATE · London

4th Estate
An imprint of HarperCollins*Publishers*
1 London Bridge Street
London SE1 9GF

www.4thestate.co.uk

HarperCollins*Publishers*
Macken House, 39/40 Mayor Street Upper
Dublin 1 D01 C9W8
Ireland

First published in Great Britain in 2023 by 4th Estate

1

Typeset in Portrait by Palimpsest Book Production Ltd, Falkirk, Stirlingshire

Printed and bound in the UK using 100% renewable electricity at
CPI Group (UK) Ltd

Contents

Alice Blanche Balfour, for inspiration and opportunity

CHAPTER 1

Getting to Know Moths

A shaded lamp and a waving blind,
And the beat of a clock from a distant floor:
On this scene enter—winged, horned, and spined—
A longlegs, a moth, and a dumbledore;
While 'mid my page there idly stands
A sleepy fly, that rubs its hands . . .

'An August Midnight', Thomas Hardy

The world of moths is full of misconceptions; assumptions made without really knowing. When did you last meet a moth? How closely did you look at it? Did you admire it, or stop to wonder how it lived its life? Whether colourful or plain, large or small, moths have lifestyles full of surprises and, even better, they can be found all around us. Their variety and beauty are there to be discovered, by anyone prepared to look.

A few years ago, on a fresh September morning, I was kneeling on the damp leafy ground of a woodland edge near my home in southeast Scotland. If I was more alert to my wider surroundings, I would have heard the distant hum of traffic from the bypass, and car doors slamming in the nearby car park as dog walkers embarked on their morning ritual. If I'd looked up, just beyond some spindly hawthorn and alder I would have seen acres of stubble field, crossed by powerlines, stretching to a line of houses in the distance. Not an obvious wildlife idyll. But at that particular moment, I was oblivious to all this.

My attention is focused on a beautiful moth, one I have never seen before. In fact, I later find out nobody had seen one in Scotland for over sixty years. In front of me is a Mallow moth.

For anyone offering this Mallow moth a cursory glance, it might seem little more than a small, dull-coloured triangle, but my gaze is more lingering. Its wings span a couple of centimetres and are layered in a medley of mouthwatering browns: a darker central stripe shaded in hazelnut, chocolate and cinnamon, sandwiched between lighter bands of milky coffee and caramel. To finish, it is sprinkled with a light frosting of sugar-white scales. I suppose brown might be a fair summary description, but it is definitely not dull.

The night before, cursing as I inadvertently veered from the path and found myself battling through a bramble thicket in the failing light, I had left a small light trap in this spot. Its bright light shone through the hours of darkness, and moths, unable to resist the allure, were side-tracked from their nightly business, falling into the box below. Here they settled in the crevices of egg boxes I had piled inside

Mallow moth

the trap, until dawn when I had returned to admire, document and release them carefully back into the undergrowth.

Pink-barred Sallow, Yellow-line Quaker, Barred Chestnut; I had already enjoyed a beautiful selection of russet, gold and tan insects this morning, their wings and furry bodies matching the palette of the falling leaves. The Mallow (*Larentia clavaria*) was the final moth, a surprise finale. I probably let out the tiniest gasp of surprise, before happiness settled across my face and I sat back on my heels to admire its understated elegance.

Moths, together with butterflies, make up a large and diverse group of insects known as Lepidoptera. Around the world, close to 160,000 different species of Lepidoptera have been described, but many more, some estimate at least the same number again, await discovery.[1]

'Lepidoptera' means scaled-wing. This is the defining feature of butterflies and moths. Their wings are uniquely covered on both sides in tiny scales; miniature overlapping flaps of chitin which under a microscope reveal an intricate architecture, reminiscent of elegantly sculpted roof tiles. A combination of pigment and the surface texture of each scale gives each wing the colour and shine that we perceive.

Insects like this have been around for a long time. Recently, tiny fossils of ancient wing scales, serendipitously found by researchers looking for pollen grains in sedimentary rocks, revealed that insects recognisable as moths were flying around well over 200 million years ago, before the earliest dinosaurs and before flowering plants.[2] These first moths were tiny, perhaps feasting on spores and pollen of pre-historic plants. Since then, natural selection has been at work, shaping their adaptations to different habitats and lifestyles as new opportunities arose. Lepidoptera are now one of the most diverse animal groups on our planet.

Distinguishing moths from butterflies on outward appearance is not always easy, but there are a few rules of thumb that can help. In most butterflies the antennae end with a small club-shaped swelling, whereas in moths they are usually feathery or taper to a fine tip. Butterflies tend to rest with their wings held shut above their body, whereas moths more usually rest with the upper sides of the wings on full view. Further distinctions can be made by looking for a tiny hook that joins fore- and hindwings, a feature only present in moths. The stereotypical differences, with moths cast as the drab, nocturnal cousins of colourful sun-loving butterflies are simply wrong. There are many brightly coloured moths and many which are active throughout the day. Equally, there are a few nocturnal butterflies and plenty that come clothed in shades of brown and grey. It's more sensible to think of them together; both butterflies *and* moths are beautiful and fascinating insects.

Taxonomically, the butterflies are in fact just one of many Lepidopteran 'clans'. For convenience, a bit like a family tree, Lepidoptera are divided up into forty-two groups called superfamilies, based on similarities in their genetics.[3] Butterflies form just one of these groups, the Papilionoidea. All the others, making up more than 90 per cent of the known species of Lepidoptera, contain the insects we refer to as moths. My friend who specialises in flies suggests butterflies would be better renamed 'butter-moths'. 'Too many things are called flies,' he grumbles.

Moths have important roles in ecosystems everywhere. As caterpillars, their nibbling helps keep plant growth in check and they provide nutritious prey for other animals. For example, one tiny blue tit chick in an English woodland can consume around a hundred caterpillars in a day. Factor that up to all the blue tit chicks in all the blue tit nests, and then consider all the other bird species with nests of chicks and you have a huge demand for caterpillars. In years when there are fewer caterpillars, fewer songbird chicks fledge.[4]

As adults, moths are not only juicy prey for other creatures, many are effective pollinators and most work the night shift when our more familiar flower-visitors such as bees and hoverflies rest. Across the countryside plants are reliant on these furry-bodied visitors to carry pollen grains from flower to flower, helping them reproduce and set seed. Recent studies in Britain found at least 289 species from 75 plant families were partially or exclusively pollinated by moths.[5] This insect revelry is easy to enjoy by shining a torch on flowers after dark. On warm spring evenings, soft willow catkins can throng with Hebrew Characters (*Orthosia gothica*) and Common Quakers (*Orthosia cerasi*), their bodies dusted yellow with pollen as they go. Later in the year flowering campions, buddleia, orchids, even ivy are among the nectar bars entertaining different clienteles of thirsty moths.

Studies from around the world have shown that moths, along with many other insects, are in trouble. Overall, their biodiversity is declining. This is worrying in itself, but changes in the balance and abundance of moth species will also have devastating consequences for other parts of their ecosystems. Through habitat destruction, climate warming and global trade, humans are without doubt responsible and most scientists agree that urgent action is required to divert catastrophe before it is too late. On an individual level, this means making lifestyle changes so that we consume less and lobbying politicians and decision-makers to implement changes for a sustainable future.

Monitoring moths is a good way to keep tabs on ecosystem health. If moths appear to be struggling, it is an indication that other plants and animals living in the same environment are also likely to be suffering. Their short life cycles – moth species complete at least one generation in a year – mean habitat degradation can quickly lead to a decline in population size. As different species of moth have different habitat requirements, changes in the abundance of individual species give clues as to where the problems are.

Moths also have another important role in conservation. They can be found everywhere, from parks and gardens in cities to the remotest of nature reserves. As insects go, many are large and colourful and easy to see up close with the help of a light trap. They are perfect ambassadors for local nature, the hidden world of wildlife on our doorsteps. As more people notice and enjoy the moths secreted all around them, this will lead to greater awareness and enjoyment of other local wildlife and green spaces. Town or countryside, old or young, moths are available to all.

My first memory of looking for moths was when I was around 8 years old. My father had suggested we have a go at sugaring. This technique, much used by the moth collectors of old, has perhaps fallen out of fashion in more recent times as the all-too-easy light trap has gained popularity. Essentially, a concoction of treacle, brown sugar and beer is heated together into a sweet sticky liquid. Once cooled, the syrup is painted onto fence posts just before dusk, luring the moths in for an energy-rich feast before they disappear back into the night. My father is a high-achieving medical scientist, back then more used to running precise laboratory experiments with whirring equipment, surrounded by white-coated colleagues. He didn't let on at the time but I'm not sure he really knew what he was doing. I suspect this foray into the dark art of moth-luring was as much a first for him as it was for me. The moths never came but I don't remember being disappointed; potion mixing with my father after bedtime was enough of an adventure.

A couple of years later, perhaps after seeing the equipment in the pages of a book borrowed from the library, I tried making my own light trap for attracting moths. I placed my bedside lamp (shade removed) into a large box and took it into the garden, with the help of as many extension cables as I could find. No parental involvement here, and certainly no risk assessment. I don't remember it being a huge success, although, unlike the sugaring occasion, I did at least catch a few moths: two Hebrew Character are all I can recall, a mostly brown moth with striking black marks on each wing which apparently resemble a letter from the Hebrew alphabet. Aged 10, the idea of alternative alphabets was itself a new discovery and the name was probably more memorable than the moth itself. If there were other moths in the trap, their identities have been forgotten or were never known.

Teenage me enjoyed a bird-centric natural history focus; speeding off on my bike before school to check out a nearby pond or adding binoculars to the textbooks in my school bag and alighting from the bus several kilometres sooner than necessary, so I could birdwatch my way home. I went on to volunteer on nature reserves around Britain and had the privilege of helping with wildlife surveys in Tanzania. Maybe it was there, among the weird and wonderful tropical creatures which I could not put a name to, that my interest in insects was truly ignited.

Zoology at university had me creeping around local allotments after dark in pursuit of New Zealand flatworms. Moths went unnoticed. I went on to work on conservation projects, pursuing an eclectic mix of creatures from earthworms to pike. Next came a PhD and postdoctoral research, using aphids and spittlebugs and some nifty laboratory techniques to study plant physiology. I worked with people researching butterflies, and gleaned random snippets of Lepidopteran biology through them, yet if I paid any particular attention to the moths flying in the countryside around me, I no longer recall it.

Fast-forward a decade or so and, with the freedom unleashed by my youngest child starting pre-school, I began volunteering with my local nature conservation group to help on wildlife reserves around Newbury. I learned about nightjars and adders; ground-nesting birds and dog-walkers; hazel coppices and ancient oaks. I made friends who had nothing to do with babies or toddlers.

One evening I joined the volunteer group for a light-trapping session on Greenham Common. I had two young daughters with me, first excited, then sensible, then fidgety. As dusk fell and the light bulbs filled the air with their shine, the moths flooded in. My adult company ranged from young conservation trainees eagerly leafing through the guidebook, trying to put a name to every mothy visitor, to a more elderly expert, a guidebook in himself, who was going to stay all night, snatching a few hours' sleep in the back of his converted van, while we youngsters headed home to bed.

We saw Shuttle-shaped Dart (*Agrotis puta*) (recognised by the impression of a weaver's shuttle on its wing) and Treble Lines (*Charanyca trigrammica*) (you've guessed it, with three parallel lines delicately striping its wings). There were many, many more, but even without putting names to them there was a pleasure in the time and the place – a warm evening, friendly folk and the flutter of insects flying in from the dusk. The moths emerged from heather tussocks, gorse bushes and birch trees, from grassy verges and moss-covered walls. From the very places I passed through almost daily when walking the dog. Yet even I, relatively observant of the wild-life around me and well aware that there are hundreds of species of moths in the UK, wasn't prepared for quite such the extraordinary variety that appeared in so short a space of time.

Not long afterwards we moved to Scotland, by which time the seeds of an obsession had been sown.

My husband had first wooed me during my PhD by impressing me with his electrical engineering skills, building equipment to detect the electrical currents made by feeding aphids. Now these talents were put to good use once again to make me a moth trap. Unlike the shabby attempts of my youth, this one was constructed of sturdy wood and Perspex, the bulb was a sky-lighting mercury vapour bulb and all electrical contacts were safely waterproofed. For the first time in my life, I began to record moths properly.

Over the weeks, the at-first amorphous groups of brown moths that turned up in the box overnight slowly became distinct and my camera's memory card filled with images of surprisingly beautiful creatures. The delicacy of the Light Emerald (*Campaea margaritaria*), large pale-green wings bisected with white, the tips dabbed with crimson; the Muslin moth (*Diaphora mendica*), a wonderful furry grey-winged moth which, when viewed face-on, reveals dazzling tangerine orange furred front legs; the Garden Tiger (*Arctia caja*), forewings a mosaic of darkest brown and white which conceal shocking scarlet underwings spotted with denim blue. Such exotic creatures living in my Scottish garden! So colourful, yet completely evading notice by day.

I shared the wonder with my children, who were not yet old enough to feign indifference. I went along to a local 'moth morning', only 10 kilometres from my house yet with a moth variety different from that of my garden. Then, in 2014, I organised a community bioblitz. For twenty-four hours the task was for everyone to record as much wildlife as possible within the confines of a local village. The primary school was involved. Wildlife experts were invited. Moths, of course, featured. I borrowed a generator and slept out with my moth trap under the stars. I hadn't done this sort of thing for years. The following morning, sleep-deprived but enthusiastic, I shared the moths with a fascinated group of early risers, my rudimentary identification skills bolstered by support from a more experienced moth expert. If there was a turning point then this was it. After that night, my occasional flirtation with moths finally settled into a long-term relationship.

I received a second trap for a birthday, this one powered by a battery which allowed me to be mobile. No longer restrained by the proximity of a three-pin plug to power my bulb, I could head out into wilder countryside, where an ever-greater variety of moths awaited my discovery. More traps were constructed. Soon I had a small fleet and I was leaving them in several different locations overnight. Moths gave me the means to enjoy a more intimate connection with my local countryside, to witness sunsets and sunrises, away from all domestic responsibilities. This time became my own valuable me-time, as I discovered that late nights and early mornings could be efficiently shoehorned between bedtime stories and school runs. I was transported back to those carefree days of my childhood, spent roaming the fields and woods on my own, with only wildlife on my mind.

Moths have become a distraction to be sought anywhere. A brief stop-off on a long car journey; a family walk with the dog; or a solo trip to the hills. Looking out for moths gets me noticing nature more closely, provides me with countless new opportunities to learn, and as an added bonus they are easy to share with friends, family and colleagues who find themselves in my company. The information I gather might be contributing to science and conservation, but the pursuit of moths is important for my own well-being too.

In this book, through stories from my own adventures getting to know moths better, I hope to share insight into their remarkable lives and inspiration to go out and discover their secret world for yourself. It's easy to do and it's fun. You might even find it becomes addictive.

CHAPTER 2

Like a Moth to a Flame

The white moths flutter about the lamp,
Enamoured with light;
And a thousand creatures softly sing
A song to the night!

'A Summer Night', Elizabeth Stoddard

With the evening meal over, it's time to load the car. The children wisely slip back to their bedrooms to avoid the bustle. Large hold-all bags each containing a light trap are retrieved from the garage. Batteries, notebook, pencil, small pots; rucksack; tent, sleeping bag, water and snacks; all pile up by the front door. The dog, on seeing the tell-tale signs, retreats to her basket. When the moth equipment appears, she knows only too well that a good walk will be ruined by the constant stopping and starting, the waiting around. Tonight she is not invited anyway. I'm on a mission, heading into the Lammermuir Hills to meet a regular moth-ing companion, Mark Cubitt.

Mark only came to moths after several decades spent as a devoted birdwatcher. In 2007 he borrowed a light trap from a fellow birder for a night, and was immediately hooked. With what I now recognise as his characteristic get-up-and-go, within a week he had made his own light trap. Three months later he had added two more traps and was out several nights a week discovering his local moths. 'I just loved their beauty and variety,' he told me. 'And I also liked the challenge of trying to improve and innovate trap designs.' He continues to enjoy this element of DIY gadgetry, for there is no such thing as a perfect light trap for moths.

As well as making them, Mark has also come up with a convenient way to carry multiple traps over rough terrain. In a modern twist on an old milkmaid's method of transporting her pails, a large bag containing a light trap is hooked onto each end of a long carrying pole. With bags attached, the pole can be hoisted up and balanced across the shoulders like a yoke, making a surprisingly easy way, for a short distance at least, of carrying bulky traps. There have been some curious stares, but only a few people have ever stopped me to ask what on earth I am doing. A rucksack is used to carry the heavy batteries, at which point the experience becomes more packhorse than milkmaid.

Our meeting place is an overgrown lay-by beneath a steeply sloping birch woodland. The trees here are thick-trunked, growing twisted and gnarled as they cling to a windswept rocky scree before it gives way to rounded heather-clad hills above, the domain of driven grouse shoots and sentinel wind-farms. Old maps show this hillside has been wooded for centuries, a remnant of what might once have been much more widespread before deer, sheep and grouse took hold. It doesn't cover a large area, but this patch of woodland is an unusual habitat in East Lothian and we are hoping to find some unusual moths living here. With six light traps between us, we will have the hillside well covered.

Mark is already there, his equipment lined up by the side of the road. At well over six foot tall, sturdily booted and appropriately hatted for the weather, Mark has become a friendly and familiar figure waiting by the roadside as dusk falls, but I wonder what others driving past might think of him, or the contents of his bags. How many would guess that moths were on the agenda?

I apologise for my usual lateness, and, with brief pleasantries as I fumble around getting my own kit readied, we get down to business. With darkness looming this is no time for dawdling anyway. Traps swinging from the pole hoisted across our shoulders, we set off to put them in position.

The steepness of the slope and its loose scree make my going slow, even more so with the weight and uneven balance of my load, but the quest to find a good spot for my traps keeps my spirits up, just to that rock there, or maybe just a little further. Like a primeval hunter setting a trap on which the family's next meal might depend, the habitat is surveyed and sized up and the trap's position chosen with care.

I have no illusions that my primitive hunting abilities are anything other than vestigial but there is an element of strategy involved in placing the traps. With several traps in one area, the aim is to choose slightly different surrounding

Mark and Katty trapping

habitat for each, in the hope that the broadest variety of moths will be attracted overnight. Some moths are generalists or wide-ranging in their habitat choices and could turn up in any trap on this wooded slope; but others are more selective, perhaps only flying under the cover of the canopy, or staying close to certain plants, or maybe more often found in places where slabs of rock warmed by the sun make the local temperature just that bit warmer. So my heaviest trap and battery is positioned first, under the canopy of a silver birch with a wood sage understorey. A few hundred metres further on and the next is perched on a prominent rock slab surrounded by a mosaic of heather and bare scree. The final one is positioned on the upper edge of the wood, with heather and blaeberry above, birch below. Satisfied, I head back down to the car, slip-sliding on ankle-twisting terrain without the weight of the traps to slow me down.

On a flat grassy area, we pitch our tents. Sunrise is about 4.20 a.m. so agreeing to get up a little in advance of that, we zip ourselves in. Sometimes on light-trapping expeditions sleep doesn't feature at all. Traps can be manned through the night with incoming moths potted up before they have a chance to escape back into the darkness. Not all moths approaching a light trap will end up in it. Consequently, catches can be considerably higher with an all-night vigil and it is a sociable way to pass the small hours. But on this occasion tomorrow's work and other commitments dictate we try at least a few sensible hours' sleep.

Dawn arrives quickly, the melody of a blackbird rendering any digital alarm clock superfluous. We pack up our tents on autopilot, as the fog of insufficient sleep lifts a little, before setting off to the traps. The rocky gradient goes unnoticed as I imagine what treasures the night might have delivered. Mark and I exchange few words; is it tempting fate to speak out loud the species we are hoping for?

Settling down either side of the first trap, we begin with a quick scout of the moths that never made it in but have settled instead on the ground around the edge. Species names are called out, like a register, and scrawled into a notebook. Familiar ones are carefully removed to the surrounding undergrowth to get on with their day; the more unusual gently coaxed into a pot for later scrutiny. Then it is into the trap itself. Peering through the clear plastic lid gives a quick preview of what's to come and any fidgeting individuals that are likely to take flight when the lid is lifted.

Mark spots it before I do. A small moth with an unusual shape is clinging to the side of the trap just beneath the lid. It is pale in colour but it is the way it holds its wings and the silhouette those wings make that give away its identity. He lifts the lid gently for a better look: Scalloped Hook-tip. Not just one, but three!

Scalloped Hook-tip (*Falcaria lacertinaria*) was one of the species we were hoping for. I had never seen one before, and although the colour and shape made it an unmistakeable match to the illustration in my guidebook, I was expecting a bigger moth. Their wavy-edged wings are a pale yellow-brown traversed by two fine lines and held tent-like over the body, a distinctive position that few other moths found in Britain adopt. The end result gives the look of a carelessly strewn withered birch leaf and I imagine would be all but impossible to spot, were they not sitting in a plastic moth trap. The last time Scalloped Hook-tip was recorded in East Lothian was in this very woodland over fifty years ago. Although found further south and further north, the moth is more or less absent from the central belt of Scotland; perhaps in this more intensively farmed and populated region, suitable habitat is in short supply. Or perhaps nobody has been looking for it. It was good to know it was still here.

As we work our way through the rest of this trap and then the others, more delightful moths are revealed: Welsh Wave (*Venusia cambrica*), True Lover's Knot (*Lycophotia porphyria*), Buff-tip (*Phalera bucephala*), Dark Tussock (*Dicallomera fascelina*), a dazzling Green Silver-lines (*Pseudoips prasinana*). Each one an imaginatively named jewel, each with a distinctive shape and pattern designed to make it blend into its usual resting place. How many more lie hidden as we record, secreted in the undergrowth and on the trees all around? We return to our potted moths, to mull over their identities and take photographs, before releasing them carefully into the undergrowth. With a brief flight or a short scuttle, they quickly blend back in to their world.

By 8 a.m., we have finished. Between us, we have 51 different species and over 500 individuals scored in our notebooks. Not bad for this habitat, in Scotland, at this time of year. With the equipment stowed away in our cars, we go our separate ways. I'm exhausted, but it's been a perfect start to the day. Memories of sitting beneath a gnarled birch tree as the early sun filters through, the shared joy of beautiful moths, the amazing dead-leaf camouflage of the Scalloped Hook-tip, keep me going through my ordinary day.

Not all species of moth are attracted to light, but as anyone who has had a peaceful evening interrupted by an erratic but persistent banging between light shade and bulb will know, most moths find light hard to ignore. It is a curious behaviour; as they fuss about the bulb they are distracted from all those things they should be doing – feeding, mating, dispersing. Worse still, it can be fatal. They waste energy reserves, damage wings and are vulnerable to predators.

We know that it is the shortest wavelengths at the ultra-violet end of the spectrum that are the most alluring. We know that single light sources are most attractive on the darkest nights. But nobody has yet discovered why moths, along with many other nocturnal insects, should be drawn to light at all. It remains a great unanswered moth mystery.

There are of course theories. One is that moths use celestial light (the moon and stars) to orientate themselves. By keeping the light source at a constant angle to their eyes they can strike a straight course, a behaviour known as transverse navigation. This could be effective if using the moon and stars as these are a long way away, but for a moth to keep the light of a nearby lamp at the same angle to its body it has to constantly turn inwards. This would cause it to spiral ever closer to the lamp, leading to an eventual collision. In reality a moth's flight pattern near a bulb isn't like this. It is more erratic. The moth might fly towards the bulb but then swerve away. It might loop round one way before zigzagging towards the light, only to change course and land nearby. Transverse navigation isn't able to explain this behaviour.

Other theories suggest a moth flying out from the dark is dazzled by the bright light and becomes disorientated, losing its ability to see properly. Or maybe the moths are actually trying to avoid the light but their eyes become confused by the contrast between the bright light and the surrounding darkness, leading to erratic and confused flight. Moths do seem to seek out darker areas in the vicinity of traps to rest, so perhaps these ideas offer at least a partial explanation.

Unsolved mystery aside, moths' habit of coming to light has long been exploited by humans eager to catch them. An early account of using light for pest control dates back to Roman times. The writer Lucius Junius Moderatus Columella in his hefty treatise on agriculture, *de re Rustica*, describes narrow-mouthed pots with a burning candle in the bottom as an effective way to lure moths, presumably wax moths, away from bee hives. Unable to escape the narrow confines of the pot, the nuisance pests are killed. Over subsequent centuries, continued efforts to control wax moths in hives and, later, various insect pests in cotton fields, led to further developments in using light traps to get rid of unwanted moths. It wasn't until well into the nineteenth century that lepidopterists began to experiment with traps designed specifically for their hobby, to attract moths they actually wanted to see. The light trap is now a central piece of equipment for moth enthusiasts.

Before convenient traps became widely available, moth-hunting naturalists exploited the allure of light in other ways. For town dwellers, a pre-dawn circuit of the street gas lamps was productive and a ladder would form part of the essential kit list. Articles in Victorian natural history magazines describe the efforts of intrepid town Lepidopterists to be first to the lights in the morning, to be assured the pick of the choicest moth specimens.[1]

Some moth collectors realised that moths could be encountered without the inconvenience of leaving their house. With carefully positioned lamps and open windows the living room itself could become a giant moth trap. It was then simply a matter of getting on with your evening, occasionally rising to net an interesting arrival that flew inside. Nowadays, where domestic harmony permits, some people use bathrooms, with their smooth walls and relatively little clutter, to the same effect.

Using a white sheet was also popular. Here a light source is suspended above or behind a hanging sheet and the entomologist loiters with intent, collecting specimens of interest as they come to rest on the sheet. This remains one of the most straightforward ways to attract moths and is the method of choice in the biodiverse tropics, where the sheet quickly gets covered by all sorts of flying night-life.

White-painted walls work in a similar way. Public buildings such as toilet blocks and shops that remain lit all night for security reasons become convenient ready-made moth-magnets. Checking a car-park toilet block near my children's primary school was once a productive part of my morning routine, my 'bog list' slowly accumulating to over twenty species of moth by the time their primary school days were done. A few years ago, a keen-eyed naturalist in my local town spotted East Lothian's first Yellow-tail moth (*Euproctis similis*), resting in white fluffy splendour on the window of a solicitor's office on the high street. Just the sort of window shopping I could enjoy.

As interest in moth collecting soared into the early twentieth century, purpose-made light traps started to gain popularity. Electrical technology was improving and eager moth-ers, keen to boost their collections, devised easier and better ways to lure their moths. Traps were designed to a similar plan, which remains the basis of light traps today: a large box with a light source on top. Where oil lamps had first been used to provide the illumination, as electricity became widely accessible, light bulbs were quickly adopted. The first commercially available traps to use mercury-vapour bulbs, which have a high ultra-violet output, were designed by the Robinson brothers in the 1950s and their design has changed very little since. A large circular tub forms the base of the trap. A lid sits on top, a bit like a cooking pot lid but with a wide downward-pointing funnel in the centre. The light bulb is held just above this central funnel, attracting moths which hopefully fly down the narrowing gap and are channelled into the tub below. Once inside, it is harder for them to navigate their way back out and they quite quickly settle down to spend the rest of the night in the cavities of cardboard egg boxes layered in the trap for this reason.

A Robinson trap topped with a mercury-vapour bulb is considered by many to be the Rolls Royce of moth-ing equipment. But bulb technology is changing and there are other effective designs, bulb and battery combinations,[2] particularly for those wanting more portable traps to take into the countryside or a slightly dimmer glow to maintain neighbourly peace.

A garden light trap left on overnight becomes occupied as you sleep. The following morning, captured moths can be examined at close quarters before being carefully returned unharmed to the undergrowth. As different moths fly at different times of the year, each trapping session attracts a different selection of species to the last. How many other natural history experiences combine such close-up wildlife viewing with such variety and convenience?

Large, mature gardens usually have the greatest range of

moths, but even in the smallest and most unassuming of spaces there are moths to be seen with the help of light. In Glasgow, entomologist Scott Shanks uses his light trap on a tiny balcony. Although it faces inwards to a courtyard and car park, he has still clocked about 190 species in ten years of trapping and he continues to attract new species each year. Colourful Elephant Hawk-moths (*Deilephila elpenor*), striking patterned Hebrew Characters (*Orthosia gothica*) and two-tone Garden Carpets (*Xanthorhoe fluctuata*) are all regulars. Not all his moth finds are in the light trap. A potted Kilmarnock willow also draws in some species. In 2019 after noticing some leaf damage, Scott found a couple of impressive Puss moth caterpillars munching away. One evening a month or so later he spotted one wandering across the balcony floor looking for somewhere to pupate. 'I don't know where it ended up,' he told me, 'but in 2020 I had an adult Puss moth on the balcony railings – so I'm hoping it was the same one.' The discovery of an empty pupal case from a Poplar Hawk-moth (*Laothoe populi*) suggests that this chunky moth probably also once grew on the willow leaves, even though Scott never noticed the sizeable caterpillar.

Regular trapping in the same place allows you to track your own moths. New species turn up, formerly familiar ones become scarce. For Scott, his balcony moth records show a worrying decline over recent years. 'I'm increasingly having nights with no moths at all,' he told me. Although there could be many reasons for this downturn, he fears that nearby brownfield sites that have been developed into blocks of flats are a big part of the problem. 'There has been lots of hard landscaping and planting of exotic shrubs. Pretty depressing when you consider all the native trees and shrubs that have been lost.'

The benefit of using light traps to monitor and record moths is clear, but there is a dark side to the hobby. In using light traps, we interrupt the normal behaviour of those moths we catch. For the night they spend in a trap, their ability to disperse, find a mate and lay eggs in a suitable place is disrupted. If a trap is used regularly in the same place, local birds soon learn to associate the trap with an easy breakfast. The best way to avoid this is to beat the early birds to it, getting up just before dawn to rescue the moths from the ground around the trap. I've heard of another approach too; apparently one enterprising trapper leaves a plastic toy snake on his trap, claiming this is an effective bird deterrent in the morning.

One night I marked all the moths I caught in my garden trap with a small blob of ink on a wing. The following night several marked moths turned up in the trap again. There were also new moths and not all of the previous night's moths made a reappearance, but even so, for some moths this was a second night in a row spent in my trap. Weather conditions, time of year and habitat will influence moths' nightly activity but it made me think about my impact on individuals and how frequently I should trap in the same place.

Paul Waring, well known among British moth-ers for his name on the spine of one of the most popular moth field guides, has investigated this more thoroughly, with many hours spent marking moths and tracking their movements, starting in the 1980s with a PhD on woodland moths. He found that consecutive nights trapping in the same place can catch as many as 15 per cent of the first night's moths again. Even with a full week's gap between trapping, a small proportion of individuals can be recaptured, though at least for these they have had the intervening nights to get on with their lives without distraction.[3]

Releasing captured moths some distance from the trap in the morning significantly reduces the chance they will turn up in the trap again the following night, but Paul's studies have also shown that some moths range very little during their lives. Moving these more sedentary species, even short distances and perhaps into a slightly different habitat, increases the level of interference. It is impossible to use a light trap without any impact on the moths they capture but Paul suggests that for minimum disturbance, moths should be released from the trap into the undergrowth where they are caught and that at least one night, ideally more, should elapse before the trap is used in the same spot again.

Of course, entomologists' light traps pale into insignificance against the pervasiveness of light pollution. Many of our cities, towns and, increasingly, rural areas are lit all night. True darkness is now a rare commodity; throughout the continent of Europe few areas are unaffected by artificial lights. It has been estimated that nearly a quarter of the earth's landmass now suffers from light pollution, and over 80 per cent of the human population is affected by this.[4]

There is little doubt that moth populations are suffering

from the impacts of climate change and habitat destruction, but could light pollution be another problem they face? A comprehensive Dutch study published in 2018[5] found that the biggest moth declines were in those species that showed a strong night-time attraction to artificial light. Significant declines were not recorded in moth species active only by day, nor for those species that tend to ignore light at night. The researchers concluded that artificial light must be a contributing factor to moth declines.

Various other studies and experiments have shown how this might happen. Bright lights can dazzle and disorientate, drawing moths away from their usual habitat or upsetting usual migration or dispersal routes. Predators such as spiders and bats learn to forage around lights and many moths are taken this way. Light affects life cycles. Laboratory experiments have shown that moth reproduction is disrupted without a daily dose of darkness – pheromone production is reduced, sperm production limited and egg laying diminished. Some studies have shown fewer moths visit flowers under artificial lighting. Even if the moths are going elsewhere for their dinner, the flowers under the lights are missing out on the important pollination services moths provide and so producing fewer seeds.[6]

However, despite convincing examples showing how individual species are disrupted by artificial light, the extent to which artificial lights at night are impacting populations of moths in the wild remains unclear. The most heavily lit areas also tend to be the most urbanised, where it is hard to isolate the effects of light from other man-made influences.

In reality, moths may become better adapted to living under their glare or simply avoid areas with lights.

Few people in such a short time have contributed as much to our understanding of British moths as Douglas Boyes. As a nature-loving boy, he was first drawn to moths at the age of 12 after seeing a light trap in action. He was really excited by the fact that you could set a light trap in the evening and the next morning 'these cool insects would be waiting for you!'

His fascination with moths endured, leading to university degrees with mothy themes. One research project saw him dissecting old birds' nests to discover the tiny moths that feed on old feathers and dung; his 'moth shed' in his garden was filled with bagged nests, as he patiently waited to see what moths emerged from them. Another investigation had him analysing vast amounts of data to see if he could work out reasons why some moths are actually becoming more common in the UK. For his PhD, he undertook ground-breaking research on the impact of streetlights on moth caterpillars living in roadside verges.

Meeting up over Zoom, Doug tells me more. Even through the computer screen on my desk his enthusiasm for moths is palpable. Gesticulating wildly, his eyes dancing and brain whirring, he imparts some of his vast moth knowledge to me. For his fieldwork he selected twenty-six stretches of roadside verge, each of which had a section that was lit all night and a section that remained unlit. He then set about collecting the caterpillars that lived in the verges, and compared catches between the different sections of each. Patrolling road edges in the dark, kitted out in high-vis safety gear, sweep net in hand, he must have stood out as unusual. Have any night-time commuters ever pulled over to enquire of his pursuits, I wondered. 'Not really,' he laughed, 'though my road cones did once hold up a police-led convoy transporting something big out of Aldermaston. The police had a few questions then.'

Four hundred hours of fieldwork and nearly 2,500 caterpillars later, he had some striking results. Lit areas had significantly fewer caterpillars. Not only that, but LED lights, so often marketed as 'eco-friendly', had a bigger negative impact on numbers than the traditional yellower sodium lamps. Doug thinks that fewer eggs were being laid under the glare of lights, either because fewer female moths chose to lay there or they laid smaller batches of eggs.

His study clearly demonstrated light has a damaging effect on moth populations in the field, but he suspected the effects were quite localised to where the lights were.

He went on to estimate the proportion of land in his study area that was illuminated by artificial lights or predominantly built up. It turned out only 7,000 hectares, less than 1.5 per cent of land area, was affected in this way. It is a crude analysis, but the majority of space in this well-populated area of southern England is out of direct glare of artificial light. Moths still have plenty of darkish places they can go. Of course, this doesn't mean that these places are good for moths. Intensive arable farmland, with its crop monoculture and chemical spraying, tends to be poor habitat regardless of how dark it is. Doug calculated that local nature reserves covered less than 1 per cent of the land area within his study area.[7]

'I think climate change and habitat loss are probably the thing we should be most worried about when it comes to declining trends in biodiversity,' Doug told me. But he was quick to point out that addressing light pollution is something we can easily do and therefore should do. 'Reducing the amount of artificial light at night can only help, and the positive impact on wildlife will take effect immediately. It will also save money and reduce carbon emissions.'

For example, homeowners can turn off security lights. Motion-sensitive lighting can be used. Streetlights can use bulbs which emit longer wavelengths of light that are less disruptive to moths and the lamps themselves can be better designed to focus the light downwards onto the road and pavements where it is needed rather than upwards and outwards into fields and hedgerows. Areas must be kept safe at night, but there is plenty of scope to use technology to improve the ways we do this. Within weeks of Doug's research results being published, some local councils had pledged to improve their lighting practices. The benefits will reach far beyond moths and their caterpillars; dark nights are important for a wide range of wildlife, and humans too.

Entomologists' light traps, used sensibly, have been crucial in improving our understanding of moth distributions but they have another important use. They provide an easy opportunity for people to enjoy and share a close-up wildlife experience, almost anywhere. One of my most rewarding experiences with a moth trap was at an old people's home. I didn't really know what to expect but on the morning, with a rather meagre offering of moths to show, I was beginning to regret the whole idea.

The three ladies (aged between 87 and 96) who joined me in the courtyard, with tartan blankets warming laps against the morning chill, showed such delight in the experience that my reservations were quickly dispelled. None of them had ever held a moth. They had no idea that such variety existed. They thought all moths ate clothes. As we watched a Large Yellow Underwing (Noctua pronuba) crawl across one woman's hands, wings quivering in readiness for take-off, there was genuine wonder in their faces. After a pause, as if to allow us one more admiring look, the moth took off with the briefest flash of its bright orange hindwings and landed somewhere in the nearest rose bush. This numerous, everyday moth had its audience enthralled.

Not all moths are active during the darkness. About a fifth of the world's species spend the night tucked up out of sight, only waking up for action when the sun rises. Among these day trippers are our most colourful moths, challenging both the drab and the nocturnal stereotypes. These species don't trouble themselves with artificial lights at night, which means we must use other approaches to catch up with them.

CHAPTER 3

In Broad Daylight

In every walk with Nature one receives far more than he seeks.

John Muir

A few kilometres from where I live lies John Muir Country
Park. It is a wide-open coastal space named after the local-
born man who was passionate about wilderness and its
importance for people and wildlife. Surfers brave the North
Sea swell waiting for the perfect wave; dog-walkers and families enjoy the vast beach; birds probe for rich pickings from
the saltmarsh and in between all this spreads an undulating
expanse of sand dune.

Stiff-stalked marram grass glistens grey-green in the ever-
present breeze, its roots forming a busy network below the
surface which helps hold the hummocks of soft shifting sand
in place. In scattered patches where the ground is firmer
other plants have taken root. Some, such as restharrow and
bird's-foot trefoil, grow by creeping, clinging close to the
sandy soil. Others like thistles and viper's bugloss stand tall,

their flowers beacons for thirsty insects. Many specialist plants and animals thrive in these seaside sandscapes. There are mosses and lichens; flies and spiders; beetles and snails; and, of course, moths. With numerous options for selecting a comfortable reclining spot out of the wind, and with the lulling repetitive splash of the waves in the background, it offers a blissful retreat for soaking up the sun and an easy place in which to forget time.

During June some of the marram stalks become decorated with small papery parcels. A couple of centimetres long and roughly cylindrical, tapering to a point at each end, they have a pale yellow-gold shine and are at first hard to spot, despite their prominent placing. But once your 'eye is in' they reveal themselves everywhere. These are the cocoons of the Six-spot Burnet moth (*Zygaena filipendulae*). In the coming weeks, within each flimsy case a moth pupa will complete its change from a stumpy green, yellow and black caterpillar to a striking scarlet and dark metallic-green winged adult.

There are many different kinds of burnet moth around the world, all snazzy-looking insects, many winged in finery of the darkest glossy green and brightest red. Some species are common and widespread, others are vanishingly rare. The New Forest Burnet (*Zygaena viciae*), Slender Scotch Burnet (*Zygaena loti*) and Mountain Burnet (*Zygaena exulans*) are among the most endangered moths that live in Britain.

The Six-spot Burnet is Britain's most widespread burnet moth. It is found in flower-rich grasslands where the caterpillar's foodplant bird's-foot trefoil grows. At peak emergence some time in July, flowers become covered with the striking adult moths as they clamour for drinking space to stock up on sweet nectar. This is a sight worth seeing and each year I try to time a trip to John Muir Country Park to enjoy it.

On some stems I see couples, joined end to end in pursuit of procreation. On others there are moths fresh from the pupa patiently waiting for their wings to fully unfurl so adult life can begin. Under the summer sun, their burnished wings gleam and their scarlet spots dazzle. Some females seem to barely have the opportunity to dry their wings before a male is in attendance, eager to mix his genes with hers and produce the next generation. My immediate thought is to feel a little sorry for these harassed females, but on reflection it is very likely that they made the first move by releasing an irresistible scent to invite males close; she will be keen to mate and lay her eggs as quickly as possible. Some years on a sunny July day as the moths congregate at nectar sources, they appear to be in plague proportions. Definitely the most obvious insect around, everywhere you look, feeding, mating or just sitting in full view. Why are there no birds flocking in to take advantage of this ready food supply?

The reason is that the burnets taste bad. Their bodies are packed with poisons that deter even the hungriest of predators. These poisons are accumulated by the caterpillars. When the Six-spot Burnet moth young munch on the leaves of trefoils, they not only get the essential nutrients for growing, they also swallow the chemicals, in this case cyanides, which the plant produces to protect itself. The caterpillars themselves have evolved ways to deal with the plant's poisons, so happily eat the leaves with impunity. They cunningly store the dangerous chemicals in small pouches just under their skin. This prevents the toxins from interfering with their own body functioning and also allows them to be exuded through the skin as foul-tasting droplets for any predator foolish enough to investigate too closely. When the caterpillar goes through metamorphosis the poison is transferred to the adult moth, where it continues its protective role.

We now know that the cyanide-containing poisons that burnet moths acquire as a caterpillar play more than just a defensive role in their lives. They are an important chemical currency, a valuable asset to possess, and if a caterpillar can't derive enough from its food directly, it manufactures its own to top up its stores. As adults, females use cyanides as part of their alluring perfume to help attract males, and mating males transfer varying amounts to the female in little packages with their sperm. A toxic male is the most attractive and desirable.

Dame Miriam Rothschild was one of Britain's greatest twentieth-century entomologists, a fascination with chemical defences in butterflies and moths just one part of her many and varied contributions to science. Despite little formal education, by the end of her 96 years she had accumulated numerous honorary degrees and authored hundreds of scientific publications.[1] From stories I've heard, videos I've seen and books I've read, it seems she was indefatigable, constantly observing and forever wondering why insects behave the way they do. She herself once admitted in her characteristic, matter-of-fact manner, 'I'm never bored. Never tired.' Her subjects of study in Lepidoptera alone ranged from pigments in wing scales, to brush organs on cervical glands, to the origin of poisons, to gardening for butterflies. Sharp observation skills, enviable intelligence and disregard for convention were key elements of her success, though her background of family wealth and contacts undoubtedly helped.

Miriam's father, Sir Charles Rothschild, was simultaneously a successful banker, a pioneer of nature conservation in the UK and a world authority on fleas. Her eccentric uncle, Lord Walter Rothschild, amassed a globally important zoological collection, including thousands of moths, which he housed at a purpose-built museum in Tring. Both men were important teachers and inspirations to young Miriam. She went on to spend much of her younger adult years furthering her late father's work on fleas, making her own significant contributions on, among other things, myxomatosis and rabbit fleas and unravelling the mechanisms of their extraordinary jumps.

However, after years peering down a microscope marvelling at the finer intricacies of fleas, she decided her eyesight needed a rest. Much of her attentions were then turned to butterflies and moths, and in particular how they acquire chemicals that make them distasteful to birds. Her subsequent discoveries, often in collaboration with other scientists, were considerable. She was the first to demonstrate that burnet moths are able to synthesise toxic compounds for themselves, as well as acquiring them directly from the plants they eat. Apparently, she even went as far as tasting one.

Chemical defence is an important weapon against predators for many moths, but for it to work best a predator needs to know the moth is foul-tasting *before* it takes a bite. A naïve predator that doesn't know better finds out the hard way and quickly learns to avoid anything else that looks similar. One or two moths might get damaged in the process, but thereafter the others escape harm.

To help a predator learn quickly, these moths need colours or shapes that are easy to recognise. The warning colours of the animal world are reds, oranges and yellows, usually striped or spotted with black. Predators rarely want to risk time studying a potential prey item in detail before deciding whether or not to attack – first impressions count and the bright contrasting colours will be that first impression. All sorts of animals, from wasps to salamanders, adopt a similar colour code to advertise their toxicity. By sharing the same colour scheme, toxic prey share the teaching responsibility and make it even easier for predators to learn to associate certain colours with foul-tasting food.

Storing poisons in your body isn't straightforward though. Not all moths bother, making them a tasty protein-filled snack. Why don't they cheat and pretend to be poisonous by wearing a bright disguise? It turns out that a few species do, though it isn't that simple.

The problem comes when, if too many tasty moths cloak themselves in warning colours, a predator learns that sometimes these colour combinations are good to eat. Then the warning message is no longer as effective, and as the moths are easy to spot, they are more likely to get gobbled up. In addition, a poisonous chemical may not be noxious to all predators, and even if it is, a predator that is desperately hungry might be prepared to risk the unpleasant side-effects of foul-tasting prey. The result is a complex balance of palatability, warning colours and a variety of other defences that are maintained by natural selection. The consequence: moths in a wide range of shapes, sizes, colours and flavours.

Rather than bold defensive colour clashes, some daytime fliers dazzle in other ways. Some moths shine in the sunlight.

Green Longhorn (*Adela reaumurella*) is a small moth, about a centimetre long. Its wings are plain but beautiful; their microscopically textured scales reflect and scatter the light, producing a dark metallic green-bronze lustre. This adult splendour is in stark contrast to their dowdy youth. The pale-coloured caterpillars live a life among leaf-litter, sheltering in a pouch made from dead leaves, feeding on dead leaves and finally pupating among dead leaves.

They emerge as splendid adults in April and May with one thing on their mind – finding a mate. To do this, the males gather around the tree tops in small swarms, hovering in an undulating, bobbing dance. Their performance is one of the great unsung wildlife spectacles in Britain.

The Green Longhorns' wings gleam gloriously as they catch the early spring sunlight, but that isn't their only remarkable feature. The antennae of the females are long, but in the male the length verges on the ridiculous – almost four times as long as his body. His eyes are impressively large too, seated in a bushy black face, affording a wide angle of view. Perhaps this allows him to keep an eye out for potential mates while maintaining his energetic dance moves.

Watching their sparkly activity in the spring sunshine, you could easily think they are merely enjoying some *joie de vivre*, but these mini swarms are serious; each male is after a female partner and hoping to get noticed. The females rest on nearby leaves, assessing the situation before flying in. It is hard to know whether it is the female who selects a male with the dance moves she likes, somehow signalling him to follow her into the foliage, or if it is the male that swoops in to make his claim on a passing female. Either way, the end result is a successful pairing and mating in the undergrowth.

I have a perfect spot to watch this springtime spectacle. On a sunny day in May I make an annual pilgrimage with a similarly entranced friend to a steeply sloping oak-rich wood-land. Our tickets are free. After first admiring the moths flying around their leaves close-up, we retreat uphill a little, nestling into prime grass tussock seats to enjoy a near-canopy-level view onto the trees below. Relaxing in the almost warm sunshine, flask of coffee in hand, we exchange pride and worry over teenagers, discuss things we've read, gossip about people we know. And in between, we watch through binoculars as the moths, belying the importance of their actions, jostle and bob as they put on their show.

★ ★ ★

Later in the year, the herbaceous borders of parks and gardens host a much larger and supremely nifty flyer to look out for. The Hummingbird Hawk-moth (*Macroglossum stellatarum*) uses its aerobatic skills both to travel long distances and access food. My first encounter with this speedy package of energy was when daydreaming away a warm summer's afternoon in a tiny courtyard garden in Norfolk. I was slowly awakened into reality with a curious rattling hum. Following the noise with my eyes led me to a whizzing blur of action, busy over blooms of red valerian growing in a large tub. Details beyond a greyish-orange blur were hard to discern and it was easy to see why Hummingbird Hawk-moths are frequently, usually somewhat excitedly, mistaken for a tiny hummingbird – despite this being one of the least likely birds to turn up in Britain.

Not only does this chunky hawk-moth hover at flowers, with its proboscis stretched out like a beak as it drinks nectar from deep in the flower head, but the scales at the tip of its abdomen form a fan shape, and look very much like the tail feathers of a bird. These tail fans help the moths manoeuvre, acting like a mini-rudder. They are regular visitors to gardens where the nectar of flowers like lavender, valerian and buddleia provides energy for its high-octane life. Their wings, just a few centimetres long, can beat an amazing 70–80 times a second, powering the moth at speeds of up to 20 kilometres per hour. Some years I catch fleeting glimpses of one in my Scottish garden, but I rarely get a lasting view and I have yet to have one obliging enough to hover about until I have got my camera at the ready.

Hummingbird Hawk-moths find Britain's cool wet winters hard to survive, so each year it is predominantly individuals migrating from southern Europe that arrive on our shores. But there are increasing records of individuals trying, and likely managing, to overwinter here in the shelter of buildings and barns, as they do further south. As global temperatures rise and British winters become milder, perhaps this will happen more regularly.

Not all day-flying moths are colourful or easy to spot. Clothed in cryptic browns and greys, some meld into their background among vegetation and rocks. Active only by day, and only taking to the wing when it is warm enough, patience and a sharp eye are needed to discover these more secretive species.

Mother Shipton (Euclidia mi) is a beautiful day-flying moth of grassy places. The wings are patterned in beautiful creams and browns with darker marks etched across them. These markings provide perfect camouflage among grass stems and, by coincidence, also depict the striking profile of an ugly face on each wing. With jutting chin, long hooked nose and dark beady eye it is hard to unsee the face once noticed. I came across my first Mother Shipton sunning itself on short, rabbit-nibbled grass by the coast. At the time I was just starting to get to grips with moths and had no idea what this one was. But a gnarly-faced woman on each wing was immediately obvious to me and it was this image that lodged in my head when I later sought the moth's identity in the guidebook.

Mother Shipton

This face pattern on its wings gives the moth its common name in Britain. Old Mother Shipton lived in Yorkshire in the sixteenth century. Many references describe her as a hideously ugly witch, but I think it's kinder to regard her as an erudite prophetess. She lived much of her life in a cave, making her living telling the fates and fortunes of those who sought out her extraordinary talents. According to legend, her more wide-reaching foresights were the Great Fire of London, the American Civil War and aeroplanes. Although the basis for these claims is more than a little doubtful, she sounds like an interesting lady. The cave she is purported to have lived in is now a tourist attraction. I hope she'd be honoured to have become immortalised in a moth.

Much rarer, restricted only to selected patches of moorland around Scotland's Cairngorm mountains, lives another cryptically beautiful day-flier: the Netted Mountain Moth (*Macaria carbonaria*). This is one of Britain's scarcest moths. Worryingly, despite increased efforts to search for it, it seems to be contracting even further in range and has not been found in many of its former haunts for over twenty years.

Netted Mountain Moth caterpillars are fussy. They only eat leaves of bearberry, a small shrubby plant of moorland and heathland, so named because, in some of the places that it grows, bears like eating its berries. The plant is patchily distributed on moors across the northern half of Scotland, yet the moth is only known from a fraction of these. Presumably additional factors such as the geology, altitude and microclimate also need to be right. Changes in moorland management, not to mention the insidious impact of climate warming, are piling the odds against their survival. Even in the small areas where they are thriving, huge obstacles (like mountains) prevent easy dispersal to new areas that may become suitable. All Scottish populations are classed as vulnerable and their future here far from certain.

For now at least, one reliable place to find Netted Mountain Moth is on the eastern limits of the Cairngorms National Park. On a warm day in early May when I was working nearby I awarded myself a lengthy lunch break to meet fellow moth enthusiast and countryside ranger Helen Rowe who, I hoped, would lead me to these enigmatic moths. If the sun shone, I was promised at least one. They had been seen here just a few days before.

I arrive ten minutes early, so I make my way along a track through the pines that shield the moor from roadside view.

The mountains of the Cairngorm plateau tantalise me on the horizon, still generously drizzled with the snowy legacy of an unusually cool spring. I emerge from the trees onto a gently undulating heathery moorland, a foot-high expanse of shrubby grey-greens spreading away in front of me. The area is flanked on all sides by birch, their freshly unfurling green leaves shining brightly in the welcome sunshine. Almost immediately, I see the unmistakeable fluttering of a small moth. As quickly as I notice it, it vanishes from sight. I have been taken unawares, ambushed before I'm ready and I have no idea what it was. As I'm probing around unsuccessfully in the undergrowth in the hope of rediscovering it, Helen arrives.

As she hurries up the path towards me, clothed in her ranger uniform of navy blue, hair escaping its ineffective clasp, I'm greeted with a cheery smile, though I notice at least half her attention is on the surrounding heather; a good tactic if you want to spot a cryptic moth. The sun is just breaking free from a cloud and we have barely got our 'How are yous?' out of the way before Helen points out a small flitting moth a few metres away.

It looks the same as the one I had seen minutes earlier, but with Helen's experience she knows immediately it is a Netted Mountain Moth. It was smaller than I was expecting, though my view is frustratingly brief before it disappears. I go to investigate where I'm sure it landed, but it is nowhere to be seen. Vanished. As if it had passed through an invisible opening in the heather to a parallel universe.

We resume our chat and start to saunter slowly across the

moor with eyes to the ground for anything that might give itself up by moving. It isn't long before a couple more Netted Mountain Moths scatter away in front of us and I begin to recognise what we are looking for. Having prepared myself for searching in vain for a rare moth, I'm pleasantly surprised at the ease we appear to be encountering them. They lift up apparently from nowhere, skimming over heather tussocks for a few short metres before hunkering back into cover. As I get more skilled at tracking their landing site, I finally get to enjoy some good sightings. It really is a small moth, delicate wings spanning less than a couple of centimetres.

The overall impression is of a dark sooty grey colour, but closer inspection reveals a complicated latticework of bands and bars in shades of black and smoky grey flecked with white. These dark-coloured scales are perfect for a moth living on a Scottish moor in spring, efficiently soaking up the intermittent heat from the sun. Today the clouds are few and the moth's heat absorption skills have worked well; they take off quickly whenever I try to sneak too close.

In the following weeks I try to find Netted Mountain Moth in similar habitats elsewhere in the Highlands. Places it hasn't been recorded; places it may have been overlooked. But the moth is small and the areas to cover large. If the sun isn't shining or my eyes are focused elsewhere it would be an easy one to miss. I have no luck.

Netted Mountain Moth adults are only around for a few weeks each year – just long enough to find a mate and lay eggs. So it isn't long before my fanciful window of opportunity to find a new site for them is up, at least for this year.

But my appetite has been whetted: the vast rugged vistas of the Cairngorms are a joy to visit; add the requirement for sun and the quest for a moth and there is a ready-made excuse to head to the mountains in spring, before the midges, and before the busy tourist season. Just as some might set their sights on a challenging Munro, or choose to visit the lapping shores of a peaceful loch, next year my own spring-time destinations may be chosen with a moth in mind.

Unlike light traps that can assemble large variety of moths as you sleep, seeking out daytime fliers requires more time and effort. Luckily this element of active pursuit is a challenge I enjoy and the reward of finding and watching these beautiful moths as they go about their daily business makes for time well spent. And if we want to understand the changing fortunes of our moths, we need to look out for all of them.

CHAPTER 4

Counting Moths

It's the little things that citizens do.
That's what will make the difference.

Wangari Maathai

My children can identify more moth species than most their age, but for the moment they choose to hide this superpower well. Biology was discarded at school at the earliest opportunity, and I sense healthy disinterest in seeing the moths their mother catches. Not that that stops me from showing them anyway. However, one summer karma was restored, when I was asked if I could provide some work experience for one of my daughter's friends who has her career sights set on something entomological.

Améline doesn't own any special wildlife recording equipment but she comes well armed with empty jam-jars and the camera on her phone. I am occasionally called upon to identify pictures she's shared with my daughter. The most memorable, many years ago, was a wonderful video of a fluffy white Puss Moth (*Cerura vinula*) she had found on the edge of the school playing field. Recently she captured East Lothian's first recorded Indian Meal Moth (*Plodia interpunctella*) in her bedroom; probably a stowaway in her guinea pigs' food. However, this week of work experience was the first time she uncovered the contents of a light trap. It was the first time she got to see so many moths, so close, at once.

Having negotiated an appropriately early start to suit a fifteen-year-old, Améline, my daughters and I arrive at the woodland where we had set three traps the previous evening. The mood is optimistic, the novelty of being together in the woods at this time of day as exciting as the prospect of moths to come.

I lift the lid off the first trap and allow the girls to turn over egg trays to discover what is there. With quiet pride, I am pleased to hear my daughters name a few of the occupants without any prompting. They must have been paying more attention to me than they let on, in their younger lives. With occasional gasps and giggles, between them the girls quickly learn to recognise more, enjoying letting the moths perch on their fingers and lifting them up to look at them eye to eye. There is plenty of variety to admire and the numbers are good without being overwhelming. But there is nothing unusual in the catch.

As we approach the final trap, that changes. Améline points out a pale moth sitting on the lid. It's one we haven't seen this morning and she is curious to learn its name. My heart misses a beat.

'Quick,' I whisper, indicating with flapping hands that they shouldn't crowd too close. 'Grab a pot!'

This is an exciting moth; one I have never seen before. I can't let it escape. With as much stealth as I can muster, which is difficult with three pairs of young eyes watching in amused anticipation, I slowly approach the moth to coax it into captivity.

It is a Peacock Moth (*Macaria notata*). An odd name, for nothing about this moth is reminiscent of the eponymous bird nor of the gaudy butterfly. It has beige-white wings spanning just a few centimetres, the lower edges jauntily angled rather than smoothly curved. There are some lovely deep chestnut brown markings bordering the forewings and a cluster of dark spots towards each wing tip which remind me of a tiny three-toed paw print. Anyone expecting something of a shimmering bird would certainly be disappointed, but I quite like its delicate pale-brown beauty. Peahen would suit it better. I later learn its Dutch name *Klaverblaadje* translates to 'cloverleaf'; an apt description of the paw-print wing marks. As well as its novelty to me, another reason for my excitement is that I know this species hasn't been recorded in this region of Scotland before. The girls dutifully admire it, but I suspect they are more entertained by my excited reaction. Possibly to them, it falls into the category of overreaction.

We continue to work through the rest of the trap, me happily muttering 'Peacock, wow!' occasionally. In the end, for the girls it is a Poplar Hawk-moth (*Laothoe populi*) that captivates them the most. They adore it. It is a large, softly

Peacock Moth

grey moth, body as thick as a finger with beautiful leafy shaped wings, large dark eyes and elegant white antennae lined with red. It poses docilely on Améline's fingers and she is delighted. When it is time to release the moth into the undergrowth it is reluctant to let go of her finger and has to be prised off one foot at a time. It more than earns its stripes this morning, as an ambassador for all moths. I expect the Peacock Moth is forgotten in admiration of this cuddly giant.

Poplar Hawk-moth is one of the UK's best-loved moths. Big, beautiful and widespread, it is perhaps this species more than any other that has swung moth doubters into moth lovers. It is impossible not to smile each time I see one, but because I see them often, they are rarely individually memorable. I only continue to remember that morning's Poplar Hawk-moth, one of perhaps fifty I saw that year, through a photo: my daughter and Améline beaming from ear to ear in the dappled morning light with the moth perched on a finger. Of the other moths we looked at that morning, without referring back to my notes, I remember none.

In contrast, I do recall each and every Peacock Moth I have seen in Scotland. There haven't been many; since that memorable first encounter with my teenage audience, just two more, both as it happens in the company of children and both spotted first by their younger eyes. I hope not a signal that my capabilities are diminishing with age.

Wildlife conservation, just like memory, often focuses on the rare or unusual. But Poplar Hawk-moths are no less important to our ecosystems because they are widespread and common. It is just as important to document them. In fact our more numerous species lend themselves well to keeping tabs on the overall health of the environment, because there are enough of them to reliably track increases or decreases in numbers.

To best protect biodiversity, we need some idea of what's there. In an ideal world this would be unbiased information about *all organisms*, from *everywhere*. This is, of course, an impossible ask, but fortunately we can do well with approximations.

First, for 'all organisms' read 'all moths'. Moths are sufficiently widespread and sensitive to environmental change to be a useful gauge for a wide variety of wildlife. Even better, a large number of species can be recorded relatively easily using light traps. If most moths are doing well in an area, most other plants and animals should be doing okay too. Second, for 'everywhere' read 'from as many places as possible'. By encouraging anyone and everyone to look out for moths and note down what they see, huge areas and time periods can be covered. I'm not sure how many professional moth trappers there are (or for that matter what the definition of 'professional' is in this context), but paid scientists could never match the output of the wider public – people sometimes known as citizen scientists – in recording moths.

Citizen science as a term has only been around for a few decades, but the concept dates back much further. Since Victorian times amateur moth recorders have been collecting valuable data which present-day scientists continue to analyse and turn into relevant information for land managers and policy makers. Although it means slightly different things to different people, citizen science is defined in the Oxford English Dictionary as 'scientific work undertaken by members of the general public, often in collaboration with professional scientists and scientific institutions'.[1]

Not everyone likes the term. For some, 'citizen' is a loaded word; 'amateur' downplays the skill or experience of many naturalists; and professional scientists or institutions need not be involved. Nonetheless, regardless of the term used, the contributions made by people recording wildlife in their spare time continue to be fundamental to understanding our natural history.

I record wildlife because I enjoy it. I like being out in the countryside, I like close encounters with wildlife, I like learning names of creatures and what they do. Through recording wildlife I've made friends and have become part of a supportive, like-minded community. As I've learnt more about moths and gained experience and confidence, I am further rewarded by being able to share my newly acquired knowledge with others. Even better, then, that as I have fun pursuing moths, the information I collect contributes to a greater knowledge of these insects in the UK and makes a huge difference to what nature conservation organisations are able to achieve. In 2021, Butterfly Conservation estimated that the work undertaken by volunteers, to protect both butterflies and moths, would cost £18 million if valued commercially.

In 2019, after a great deal of work, the *Atlas of Britain and Ireland's Larger Moths* was published.[2] It is a fabulous book. There are maps, pictures, graphs, trends, which show the distributions, fates and fortunes of all 893 species of larger moths recorded in Britain and Ireland. I can see at a glance if a moth has been found in my area or, if not, how close it has come. It shows me that Peacock Moth hadn't been recorded anywhere near East Lothian before my sighting, but is gradually expanding its range in regions to the north and south of me. Of the Poplar Hawk-moth it shows it is widespread across the country, the most widespread of our hawk-moths. I love daydreaming my way through the book's pages with my morning coffee, reminding myself of previous finds and eyeing up the moths that I might chance upon next.

The data used in the atlas came from 25 million records of moths, the vast majority found and recorded by thousands of naturalists like me in their free time. Up to the end of 2016, at which point the information was collated for the book, everyone who saw a moth in Britain and Ireland and sent the details of that sighting to Butterfly Conservation or Moths Ireland contributed. The oldest record in there, extracted from archives, is a Kentish Glory (*Endromis versicolora*) found in 1741. I'm not sure what the most recent record was, but the final observation I contributed was late on Hogmanay 2016 when I discovered a new site for overwintering Heralds (*Scoliopteryx libatrix*) in a culvert near my home. Who knows if anybody else was avoiding more public celebrations in favour of looking for moths that New Year's Eve, but I like to think my Heralds were the last moths of all to make the cut.

Collating observations from people all over the country requires coordination and organisation. In the UK, all moth records are first sent to the relevant county moth recorder. These are expert volunteers whose role is to check that the records made in their county are accurate, before rubber-stamping them and submitting them to a huge central database: the National Moth Recording Scheme run by Butterfly Conservation, one of the largest databases of insect observations in the world.

In charge of coordinating the network of county moth recorders, motivating moth-ers to send in their records and generally helping to make sure moth observations end up in the right place, is Zoë Randle. With a broad grin, seemingly boundless energy and admirable diplomacy for dealing with the idiosyncrasies of moth recorders, Zoë is an advocate of all things mothy. She loves their wonderful colours, shapes and forms and she loves how accessible they are to people.

'Set a light trap in a back garden on a warm summer's night, and beautiful insects you had no idea are sharing your space will come!', she remarks with genuine excitement.

One of her favourites is the 'gorgeous marshmallow pink and green' Elephant Hawk-moth (Deilephila elpenor). In fact, it is the dazzling colours and amazing shapes that she loves most about moths, insect eye-candy living in the under-growth on our doorsteps. Even better that these are insects with a reputation for being anything other than colourful. Zoë later confesses she sometimes feels a bit of a moth-er fraud, because she finds some of the small, nondescript brown species so much harder to get excited by.

I first met Zoë at a meeting of Scottish butterfly and moth recorders where she gave an upbeat talk about the work she and others were doing to prepare for the publication of the Moth Atlas. There were parts of the country that had never been visited by moth recorders and we knew little of the moths that might live there. Appealing to the latent compet-itiveness of many in the lecture theatre, she urged us all to gather our moth traps and head into the wilderness to these places. Who would help fill the gaps?

My family would have been unimpressed if during the summer I'd abandoned them for the southernmost tip of

Kintyre or the middle of Sutherland's flow country where there was clearly an absence of moth records. It might have led to mutiny if, worse still, I'd forced them to come moth trapping with me.

Mark Cubitt, in his capacity as my county moth recorder, helped. With his encouragement I focused my efforts on filling some of my local knowledge gaps. Mark produced a list of all the moths that had so far been seen in East Lothian and from which areas they had been found. Skimming through, I could easily pick out places lacking records, and there were plenty of gaps, even of common species. I like a focus and here was a mission: to make sure moth recording coverage in my area was as good as it could be for the forthcoming atlas.

With my sights set on putting new moth records on the map, I started to take my moth traps into the wilds of East Lothian. Leaving the family to the warm comfort of their beds, I liberated myself in beautiful corners of the county I never knew existed, peaceful pockets away from busy beaches and well-trodden paths. What better way to enjoy these than at dawn in the company of fine moths? Looking back, this was a key stage in my path to moth devotion. Collecting data for the atlas turned me from an occasional garden trapper to a many-times a week mobile trapper.

The end of 2016 cut-off for records to be included in the moth atlas was less than a year away so I didn't have much time, but whenever the weather was good, I distributed my light traps in the woods, the hills or along the coast, drawing in the moths with their brightness. Gradually the East Lothian blanks began to be filled in and, with them, my experience and skills in moth identification flourished. Those sleepless nights with infants, that I was just starting to forget, effortlessly morphed into sleepless nights with moths. Making it even more worthwhile was the knowledge that the information I was gathering would be put to good use. My records might not be of much value alone, but together with millions of others they became important.

As well as building up a comprehensive picture of which moths live where, the National Moth Recording Scheme also holds information on their changing numbers. This is a much harder thing to measure, as moths are hard to count meaningfully. The number caught in a trap depends on the weather, the type of trap and where it is placed. On warm muggy nights, moths are more active and numbers coming to light traps are significantly higher than on cool or windy nights. A large trap with a dazzling mercury vapour bulb will lure in more moths than a small battery-powered bulb. Moth numbers fluctuate naturally too. There are good years and bad years. To pick up changes in moth abundance a standardised approach is needed, ideally using the same equipment in the same places, over a time period of decades.

This is where the Rothamsted Insect Survey comes in.[3] For more than fifty years identical light traps have been run every night at set locations across Britain and Ireland and their catches identified and tallied, in large part by volunteers. With the help of clever analysis, the constant sampling effort provides information on changes in abundance.

* * *

In 2021 Butterfly Conservation produced an important report: *The State of Britain's Larger Moths*.[4] Analysing data from the National Moth Recording Scheme and the Rothamsted Insect Survey, it looked at the changing fortunes of British moths over the last fifty years, a period that happens to coincide more or less with my own time on earth. The headline figures were depressing. Overall, the abundance of British moths has declined by a third in those years. For every three moths that flew past the window as baby me slept in my cot, there are now only two. A trap in my childhood garden that might once have caught thirty moths would now only catch twenty.

Looking in more detail, it isn't all downward trends. Since 1900 one hundred and thirty-seven new moth species have become established in the UK. Surely this is good; our moth biodiversity is increasing. But it isn't that simple.

A few species have turned up unwittingly, innocent stowaways on imported plants from far-flung shores. Many arriving this way probably perish soon afterwards and are never known about, but others manage to survive and go on to spread through our countryside. Usually, these arrivals slot into their new ecosystems without fuss, but not always. One example is Box-tree Moth (*Cydalima perspectalis*), a native to Southeast Asia that was introduced by accident to Kent in 2007 and has since spread north and west rapidly. It was first reported in Scotland in 2018.

Adult Box Moth are beautiful; the commonest colour form has delicate pearly white wings bordered in black. But beauty holds no sway when it comes to damaging the plants we love. When numbers are high, their caterpillars make quick work of box bushes, disfiguring elegant topiary and sometimes completely stripping leaves from hedges. Those wanting to rid themselves of the pestilence are advised to pick caterpillars off, one by one, and squish them. If that is impractical then many resort to pesticides.

Many more newcomers get here under their own steam. Milder winters and longer springs and summers are enabling some moths to thrive ever further north, and these have become established in areas previously inhospitable. This northward spread is rapid, on average 5 kilometres per year according to *The State of Britain's Larger Moths*. 'On average'; that means there are some species advancing more quickly.

One fairly recent addition to Scotland's moth fauna is Buff Arches (*Habrosyne pyritoides*), a spectacular moth with an unusual pattern that makes it look, to me at least, as though it has a big bald patch behind its head. Closer inspection reveals no baldness; the smooth-looking patch is simply an area of silvery grey scales which contrasts with rich chestnut-brown wings patterned with a deeply undulating series of fine lines. The addition of dark to pale gradations gives a further impression of depth and texture that is hard to describe, a shapeshifting pattern that makes it at once like a moth and not like a moth.

I found my first Buff Arches nestled in a trap on the edge of a local saltmarsh in 2019. The trap was located in the hope of some tiny, nondescript, saltmarsh specialists, so my flamboyant discovery couldn't have been more off-target.

That year there were ripples of excitement as other moth recorders in southern Scotland got to enjoy this good-looking moth on their own patches for the first time, and Scottish sightings have been accumulating and creeping further northward ever since. However, this doesn't necessarily mean an overall increase in abundance. Indeed, the data show that although Buff Arches is expanding into new areas in Britain, its numbers aren't increasing. Like a thick knob of butter on toast, as they spread in range they are becoming more thinly distributed. Nobody knows why. We can only hope that as they become better established in new areas, their numbers will swell.

Against these advances, fifty-one species of moth have become extinct in Britain in the last century and many more show worrying signs of decline. As some species take advantage of milder winters to expand in Britain, for those that like some cold and dry in their lives, climate warming is not good news. The range of the Grey Mountain Carpet (*Entephria caesiata*), a moth as grey as the stony mountains it likes to hang out in, has shrunk alarmingly in Britain in recent years, as the species is pushed ever northwards and higher up hills to find the cooler conditions it likes. The strikingly beautiful Garden Tiger (*Arctia caja*) was once a common moth, the caterpillars found 'on almost every plant in almost every garden' according to the eminent Edward Newman, writing at the end of the nineteenth century.[5] This moth has decreased in abundance by nearly 90 per cent in the last fifty years and it is becoming an increasingly rare treat in light traps in the south.

Most of the data used to understand these changing moth fortunes comes from devoted enthusiasts regularly setting light traps and eager to embrace the challenge of identifying all the species they catch. The information gathered is vast but such dedication isn't for everyone. Luckily the scope of citizen science is more expansive than this. Occasional observations are also valuable, and to encourage participation by a wide range of people, there are more specific projects that have no need for special equipment or identification skills, and only a little investment in time.

One success has been the Conker Tree Science Project, launched 2009 by ecologists Michael Pocock and Darren Evans.

For Michael, and unsurprisingly everyone else I know working in ecology or conservation, spending time in nature is also an important part of his personal life. Birds are his favourite group and on the shelves behind his desk are photos of his children as toddlers, peering down his telescope on expeditions with Dad to see a rare bird; though whether they are actually seeing the said bird is another matter. As we exchange experiences of the parental joys of sharing a moth trap with youngsters – much easier to show them a moth in an egg box than a Marmora's warbler down a telescope – we find much in common. Like mine, his now teenage children are unusually knowledgeable about the wildlife they see around them, but also like my children, for the moment it is a skill they hold with a healthy mixture of embarrassment and pride.

Horse Chestnut Leaf-miner (*Cameraria ohridella*) is a beautiful moth. Little larger than a grain of rice, when viewed with a magnifier its wings are a glorious chestnut brown striped with pearly white bands edged in black. It arrived in Britain, probably on its own accord from continental Europe, in 2002. All unremarkable, except that its caterpillars cause significant damage to horse chestnut leaves and the diminutive adult moth has incredible powers of dispersal. On infested trees the leaves first develop unsightly pale tracks where the caterpillars have been chomping, then they turn brown and fall to the ground before there is even a sniff of autumn in the air.

Although it is now clear the caterpillar's mining isn't terminal to the tree, it can weaken them, and for those that love the large leafy conker trees in our towns and parklands, the caterpillars render them an unsightly mess.

Michael and Darren's Conker Tree Project was designed to encourage a new audience to get involved in noticing and recording wildlife, while also usefully tracking the spread of this tiny moth. Launched in a Bristol shopping centre, bag-laden shoppers were encouraged to take a horse chestnut leaf complete with munching caterpillars home with them in a pot, wait for the caterpillars to pupate and see what happened next. Did a moth emerge from the pupa or was it killed by a parasitoid wasp (tiny insects we'll meet in Chapter 9)?

Such was the success of that first summer, the project was soon rolled out more widely across Britain. The questions being asked were simple. How quickly was the moth spreading, and were natural enemies of the moths, in this case parasitoid wasps, doing anything to slow their spread?

A few years and several thousand participants later, there were some answers. The moths had spread quickly; each year infected horse chestnut trees were discovered ever further north and west. But the parasitoids were not very good at keeping up and ineffective at slowing the moths' spread. Michael and Darren could never have collected such a comprehensive amount of information alone; the success of projects like this relies on curious people giving it a go.

In 2015, Horse Chestnut Leaf-miner reached Scotland; four years later it was found about eight kilometres from where I live by an experienced moth-er having some downtime from granddad duties while visiting his daughter. Each summer I check the horse chestnuts lining my local park, hoping to spot some brown tracks in the leaves. I probably shouldn't look forward to finding a potentially tree-damaging insect in my neighbourhood, but it is a beautiful moth and it's only making the most of the environment it finds itself in.

The Conker Tree Project demonstrates well that monitoring caterpillars is just as relevant as recording adult moths. Only a tiny fraction of the information gathered in the National Moth Recording Scheme comes from caterpillar sightings; for Scotland it is less than 1 per cent of all records. These pesky leaf-eaters might get noticed with irritation by gardeners and farmers, but they are frequently ignored by naturalists, deemed too hard to identify. Who wants to go scrabbling in the undergrowth for soft squishy things anyway, particularly when the beautiful adults come so readily to light? But most moths spend far more of their life as caterpillars and these youngsters are every bit as charming and fascinating as the winged adult they become.

CHAPTER 5

Caterpillar Hunt

No toad spy you,
Hovering bird of prey pass by you;
Spin and die,
To live again a butterfly.

'Caterpillar', Christina Rossetti

I am perched halfway up a cliff enjoying some late August sunshine. Although my excuse for coming here was to check for moths in the tideline cave below, I was also craving the head-clearing therapy that these wave-bashed beaches and cliffs provide so well. Rough, sheep-grazed fields bounded by drystone walls give way, in places with dramatic suddenness, to the rock-strewn shores and their relentlessly crashing waves some seventy metres below. It feels far more remote than it is, and it's a place that offers space for deep breathing and relief from the invisible responsibility for the smooth running of others' lives.

Gazing without focus, the slaty greyness of the North Sea floods my view from boulder-strewn beach to gently curving horizon. The noise of waves, wind and birds is loud, but unpolluted by hums of transport or piercings of human voice. I have never met anyone on these cliffs. On occasion I've spied the sticklike silhouettes of cliff-top walkers, making the geological pilgrimage to Siccar Point. Once, when I must have been unknowingly visible, a farmer on a quadbike in the field above shouted me out of my reverie to check I was okay. But mostly I can disappear into myself in the knowledge that I won't see anyone, and nobody will see me. For a luxurious hour or two I can simply be.

This stretch of coast has a history of providing scientific inspiration. Here, in the late eighteenth century, the pioneering Scottish geologist James Hutton gathered the final evidence he needed for his groundbreaking theory that the earth is actually very old. Observing Siccar Point from a small wave-rocked boat, he realised the patterns formed where different rock types met could only have been achieved through a sequence of folding, eroding and deposition of rock over an enormous interval of time.

Hutton's ideas changed forever how people thought about the earth and its origins. In the subsequent publication he concluded, '. . . we find no vestige of a beginning, no prospect of an end'; perfectly summing up the mood of this place for me.

Having checked the dark cave (no moths), I'm making a slow ascent back up the cliffs. After many visits, I know the

Cliffs near Siccar Point

best route to forge, zigzagging between sections of vertical rock and avoiding expanses of steep, slippery scree.

Recovering my breath on a regular view-drinking pause, I marvel at the acrobatically swerving house martins which nest under tiny ledges on a sheer section of cliff. Endlessly busy catching flying insects, these birds are a familiar but sadly declining sight in small towns and villages where they like to nest beneath the eaves of buildings. Cliffs were once their original stronghold. Relaxing into my grassy perch, I enjoy fantastic views as they swoop past at eye level. My thoughts flit too, from mundane *what shall we have for dinner?* to slightly more cerebral recollections of conversations with fellow moth enthusiasts: *remember to search dunes by torchlight in June; look around sunny birch trees for Orange Underwings in April.*

Lost in pondering, idly fingering the greenness around me, a connection is made. Growing in scattered profusion is a small plant with clusters of tiny white flowers held aloft on narrow stalks branching from the main stem. Through my increasing curiosity about the lives of moths, my plant knowledge is also improving. This looks like burnet saxifrage, the larval foodplant of a small moth, the Pimpinel Pug (*Eupithecia pimpinellata*).

A more careful look reveals some which are no longer flowering, the flowers replaced with shiny reddish seed cases the size of small peppercorns. It is on these energy-rich seeds that the pug caterpillars feed. A promising sign, and with renewed purpose and allowing myself a tiny bit of excited anticipation, I proceed upwards.

This time progress is even slower and more convoluted as I scramble from one burnet saxifrage plant to another, clinging on to tussocks and wedging my knees onto ledges to carefully examine each plant in turn. A moth hunter's success is never guaranteed, but today my efforts are eventually rewarded with a likely-looking caterpillar. Small and green, with a pinkish tinge, it is a perfect colour match to the plant stem it is on.

Pimpinel Pug is just one of many species of pug, a large group of small moths which are notorious among moth recorders as tricky to identify, both as adults and caterpillars. The size, the foodplant and the habitat of the creature in front of me all fitted Pimpinel Pug, but, to be sure, experts later advised I should rear it to adulthood. Spurred on by the prospect of discovery, in the following weeks, on other sections of the coast I found more caterpillars. I kept a few in repurposed ice-cream tubs with burnet saxifrage seeds until they turned into pupae a few weeks later.

Pimpinel Pug spends the winter as a pupa, biding time in this stage so that new adults emerge at the right time of year to mate and lay eggs when burnet saxifrage plants are flowering. Knowing I had about nine months to wait, I packaged the tiny pupae into tissue-lined boxes and left them to it in the winter coolness of our garage, adorned with a clear label, 'MUM'S MOTHS!' to warn family members off. I occasionally remembered to look in on them, but it wasn't until the following June that a noticeable change began. The pupae started to darken, first at the knobbles of the eyes, then along the snaking line marking the antennae. A few weeks later, over the course of several days, each one split open to reveal a new adult moth, at last confirming their identity: Pimpinel Pug.

Fresh from the pupa it is a beautiful moth, with narrow outstretched wings flecked in more shades of tan and brown than I knew existed, quite different from the wing-worn individuals I had previously had in light traps. Excitingly, my cliff-side antics had proved the moth is not quite as rare in Scotland as previously believed.

Hunting for caterpillars can be time-consuming, but those species that specialise on a particular foodplant growing in particular places are often easier to find than the winged insects they become. There is the added benefit that cater-pillars are more stay-at-home than the restless adults, so finding one munching on a plant is good evidence that the species is reliant on that habitat. When considering moth conservation and how to manage the landscapes they live in, this is valuable information. We know surprisingly little about caterpillar lifestyles, their preferred habitats and the range of food they eat, even for many of the commoner species in the UK.

Looking for caterpillars is a particularly effective way to find our smaller moths, sometimes referred to as micro-moths. The number of different species, their diminutive size and often bewildering similarity, means that as a group they provide more of an identification challenge than their larger macro-moth cousins. Many moth recorders overlook them, or at least spend several years honing their skills on larger species first[1].

Some of the tiniest micro-moths have correspondingly miniature caterpillars that live their lives within leaves. Here they munch their way through the spongy cells of the leaf's interior, leaving the surface layers intact. Within the leaf they are sheltered from the elements, avoid the surface cells where plants store most of their poisons and are shielded from predators like spiders and beetles. Though browsing herbivores must inadvertently munch their way through many.

These caterpillars aren't much to look at, but as they eat, they move forward to reach fresh food and leave a characteristic trail in their wake. This track is known as a leaf mine and the caterpillars that make them referred to as leaf-miners. Some excavate a straight path; others take a more circuitous route. The track might stick to the leaf veins or strike out adventurously on an independent course. Their

patterns become entomological art forms; elegant hiero-
glyphs to marvel at, even without trying to work out what
they tell of the creature that made them.

For those ready for a challenge, learning to read these
foliar tracks and trails can identify the caterpillars that made
them. It is perhaps an acquired moth-hunting skill, but a
surprisingly rewarding one.

I knew Nigel Voaden by reputation long before I met him in
person; the numbers and variety of moths he reports from
his traps are extraordinary. Like many moth enthusiasts,
Nigel's natural history passion began with birds. With a
geology degree under his belt, his first job took him to equa-
torial Africa and with it, the opportunity to fill his spare time
seeking tropical birds. A decade and a lengthy bird list later,
as opportunities to find new bird species in the area were
dwindling, he started to look more closely at the moths
attracted to the security lights around the mine where he
worked. Although he'd been seeing them there for years, he
hadn't paid them much attention before. Very quickly the
shapes, colours and variety translated into a new and irresistible wildlife challenge, from which he hasn't looked back.

Moth-ing in the tropics is a different ball game to that
within the UK. With many more species, plenty still to be
described and named, identifying what you find can be a
challenge best left to experienced taxonomists. It could have
easily become overwhelming, but Nigel's approach was to
focus on identifying those he could and simply admire the
variety and beauty of the others.

Not long after this tropical moth baptism by fire, he returned to the UK and began regular moth trapping with a much more manageable array of species, yet still impressive results. Within four years he had found 85 per cent of the moths known in his home county of Fife; some 800-plus species.

'Nigel is 90 per cent moth', his partner Carla tells me by way of explanation.

He certainly puts in effort. If the weather is good, he might have six or seven traps distributed in the countryside, as well as a couple in his garden. To maximise the return for his nocturnal efforts, he has added modifications to his fleet of traps. Gaffer tape plugs up tiny gaps, a smaller funnel entrance into the trap limits the chance of moths finding their way back out, and in the morning, he zips himself and his trap into a mosquito net, so that any flighty moths trying to get away without being noticed are trapped with him, against the netting, where they can be potted and identified.

Ever eager to encounter more species, in the last couple of years he has been getting to know leaf-mining moths better. 'There is something quite addictive about looking for leaf mines that is difficult to explain,' Nigel told me. 'And there is a small but growing fraternity of recorders who are hopelessly hooked,' he added.

Autumn and winter are good times to look out for many of these species, and also a time when other moth-ing prospects dwindle. It seemed a benign, even healthy, winter addiction to have. Wanting to know more of this method of sleuthing for moths, Nigel offered to take me 'minesweeping'.

This was to be nothing like the time-wasting computer game of my youth, he quickly clarified – minesweeping for moths is to go out to a new site and make an inventory of all the leaf mines you can find. For a leaf-miner novice this sounded perfect. We made a date. I was advised to bring a hand lens, some zip-lock plastic food bags and a camera.

Meeting at a local patch of mixed woodland, within minutes we have our first species; a mine in the leaf of a beech tree just a few paces from where we park the car. The mining caterpillar has left a brown serpentine path through the dark green leaf. Holding the leaf skyward to the light, its track becomes easier to see. Nigel points out how the mine starts as a narrow path in the middle of the leaf, gradually widening as it snakes its way between the leaf veins towards the leaf edge. This was a Small Beech Pygmy (*Stigmella tityrella*).

We saunter a short distance to a birch tree and begin examining its leaves; two more mining moths are quickly identified from marks within them. As the afternoon progresses, we don't cover much distance, but my notebook, my camera's memory card and my head fill. As well as winding circuitous tracks, there are caterpillars that leave the leaf with papery brown blisters; others distort the blade into a pucker; some envelop themselves by folding a leaf edge over, or somehow manage to roll it up, securing the roll in place with minute bands of silk.

Using a hand lens to peer at the details of the mines reveals more clues of their maker. I start to recognise different patterns in the frass. Frass is the technical term for caterpillar droppings, and as the caterpillar moves forwards eating, the colour and pattern left by the expelled waste – green or brown, arced or linear – can help with identification. Sometimes I see a caterpillar still in residence. A tiny lozenge with a slightly darkened head. If I wanted, I could take these tenanted mines home and wait for the moth to complete its life cycle. It would eventually reveal itself as an adult, probably next spring.

Wandering about concentrating on leaves was unexpectedly enjoyable. No broad sweeping views here; instead the focus is on the detail of a bunch of leaves at arms' length. 'Mindfulness', but with an added dimension; in the process I learned names and discovered lifestyles of moths I previously knew nothing about.

In the end, Nigel and I identify twenty species of moth, all through the evidence left behind by their feeding caterpillars. Not a bad tally, though Nigel admits he had hoped we would find more. Eager to tap into his enthusiasm for moths further, I suggest maybe we could meet for some light trapping next year.

'That would be great,' Nigel replies. 'But be warned, I don't do quiet nights!'

In the following months I practise my mines, trying hard to remember all Nigel had shown me before it falls out of my head. What could have been sociable walks with family become punctuated with pauses as I tousle vegetation in search of signs. Sheltering from a downpour under a large hawthorn, I discover brown folded-over leaf lobes. Here the caterpillar stage of *Parornix anglicella* once lived, nibbling at less than one square centimetre of leaf before it had its fill and moulted into a pupa within its fold. Later, clearing some over-intrusive rosebay willowherb from the edge of our driveway, I notice tracks of *Mompha raschkiella*, a looping brown passage ending in a wider blotch.

As autumn progresses, I also discover green islands, a fascinating moth-bacteria phenomenon. One of the joys of autumn is the turning of green leaves to orange, yellows and browns. This is the tree's way of reclaiming some of the useful nutrients in its leaves before it loses them to the woodland floor. Sugars and starch are broken down and ferried into the branches and trunk to be put in reserve for next year. In the process, the leafy greens become browns; not good news for leaf-eating caterpillars that want the nutrients for themselves. This is where bacteria living inside the caterpillars come in. They produce a chemical which is released in the caterpillar's saliva and halts the leaf's normal ageing. This keeps the patch around the caterpillar young, green and tasty. As the rest of the leaf turns brown, the caterpillar can continue to feed in its green island until it is ready to pupate.

Soon I can't walk past any plant without casting a quick glance over its leaves. I sense a growing addiction. Waiting for my daughter outside a concert hall in Glasgow, I find myself filling the minutes by studying the meagre shrubbery in a pavement-bound planter. 'Yes, you did look a bit odd,' my daughter later informs me.

Caterpillars live a gastronomic life; this part of a moth's life is all about eating and putting on weight. Hatching from an egg at most a few millimetres long, they must pile on the pounds to reach full size, periodically moulting into a bigger-sized skin. Depending on the species, they spend from a couple of weeks to many years as a caterpillar; almost always a longer period of time than they spend as a reproducing adult moth.

Eric Carle's book *The Very Hungry Caterpillar* was a favourite of my children, its familiarity becoming such that I would read it out loud to them as they snuggled against me, sticky fingers poking the holes in its pages, while my mind was entirely elsewhere. Who knows how much entomological knowledge was imparted in the endless retelling; probably not much, for that wasn't really its aim and surely even the youngest of children soon realise that caterpillars don't actually eat slices of chocolate cake or sausages?

The book is correct though in that not all caterpillars are confined to a leafy diet. Some have carnivorous tendencies, with several species that regularly eat their siblings and a few that do away with their mother. Others chew on lichens, or fungi, or wood. In Hawaii, the caterpillars of some pug moths snatch passing flies using spiny legs and a lightning-fast grab.

Perhaps the caterpillars with the weirdest tastes of all are in the family of micro-moths known as *Tineidae*. This group contains hundreds of species and generalisations covering their diet choices are hard to make, but most feed on organic waste, performing a valuable and essential service in nature by breaking down and recycling nutrients back into the environment.

Some Tineid caterpillars specialise in digesting keratin, a

protein found in animal hair, skin, feathers, claws and horns. As they nibble, they slowly do their bit to break the tough material down. Unless you are in the habit of poking around dead animals or sifting through droppings (obviously some entomologists are), these detritus-eating moths are rarely seen. There are exceptions though. Two species have become rather well known, tarnishing all moths with their destructive reputation.

The Webbing Clothes Moth (*Tineola bisselliella*) and the Case-making Clothes Moth (*Tinea pellionella*) are problem pests worldwide. Their caterpillars feast on feathers and fur and wool. Clothes, carpets and furnishings that contain these natural fibres are readily consumed, and in the dark corners of warm houses they will happily breed all year round.

Adult clothes moths don't feed so they themselves cause no damage, but each female will lay about fifty tiny eggs on suitable substrates, which in turn hatch into the fabric-destroying caterpillars. Getting rid of an infestation is difficult. Plenty of vacuuming of small corners and crevices, lavender-impregnated capsules or sticky traps laced with pheromones are all weapons in the battle. Some suggest offering a sacrificial ball of pure wool to divert them, but the battle is usually ongoing and human victories short-lived. Constant vigilance and keeping vulnerable clothes you value in sealed bags is the only certain way to avoid damage.

Perhaps luckily, my household has so far escaped the attentions of these moths; in fact, I had never seen either until a recent stay in a hotel in the Highlands. Not wanting to ignore the prospects of a mild night in a different part of the country, but without a light trap or much energy after a day tramping over moorland, I thought I'd instead try tempting some outside moths inside. I turned all the room lights on, left the window and curtains open, and retired to the bed. Just five minutes later I was rewarded with a small moth flitting round the bedside lamp. My initial excitement was downgraded slightly when I realised it was *Tineola bisselliella* and my thoughts turned swiftly to the safety of my socks. It was the only moth I saw that evening.

Although their main aim in life is to feed up in readiness for adulthood, caterpillars provide a nutritious snack for many other animals and must devote at least some of their attention to avoiding being eaten. Just as with some adult moths, being poisonous is one strategy. The striking orange and black stripes of the Cinnabar Moth (*Tyria jacobaeae*) caterpillars make them easy to see on the ragwort plants they feed on, but birds soon learn what such colours advertise and leave the foul-tasting caterpillars alone.

Being hirsute is another caterpillar defence. Some species have hairs containing irritating toxins, but more often it is just the length or bristliness of the hairs that makes them more than the mouthful of tastiness a bird might have bargained for.

For most caterpillars though, it is a case of relying on good old camouflage to avoid detection, which explains why vegetation-matching brown and green colours are so common. Particularly impressive are those that pretend to be a twig. These not only mimic fissures and buds with knobbles and bumps on their skin, but when threatened they stretch out at an angle from their branch in a convincing twig-like pose.

I once unwittingly acquired a Brimstone Moth (*Opisthograptis luteolata*) caterpillar. It was discovered crawling up my arm after an off-piste woodland walk. Greyish-brown with a bud-like bump halfway along its back, I doubt I would have spotted it in the wild; presumably I had dislodged it from one of the bushes I'd been battling through.

I provided it with a leafy branch-filled enclosure on my desk, and over the subsequent few weeks it engaged me in many distracted minutes of playing spot-the-caterpillar. So good was this caterpillar's masquerade that my husband, admittedly not renowned for his powers of perception, was unable to recognise it as a living thing, even when I was pointing directly at it. Unfortunately for Brimstone moths, hungry birds are more observant, but clearly enough get away with their stick ruse to reach adulthood and reproduce. The bright lemon-yellow moths are common in gardens, hedge-rows and wooded places throughout Britain.

For those without any special mimicry skills, avoiding the beady eyes of birds is helped by hunkering down deep within vegetation during the day, only crawling out to feed on leaves at night when birds are roosting. Night-time searching by torchlight is a good way to discover these species, and this was how, in June 2020 with retrospectively remarkable ease, I made my most significant caterpillar discovery yet.

Mallow moth (*Larentia clavaria*) was thought to have become extinct in Scotland until one had quite unexpectedly and excitingly arrived in my light trap the previous autumn. To investigate if the moth really was living here, or if mine was simply a one-off wanderer from afar, I wanted to search the area for its caterpillars. They feed on species of mallow and hollyhock, so it was these plants that I was trying to remember to keep a look-out for when out and about.

I had spent the morning with a friend, enjoyably but rather fruitlessly exploring a local woodland for day-flying moths. My friend finds flies more interesting, so was looking for them. As usual these were enviably numerous, but the down-side is that most are fiddlier to identify than moths, and he would have to wait until he could examine each under a microscope to know what he had caught. In the dappled shade of lunch, I had been telling him about Mallow moths and the failure of my quest to locate any hollyhocks or common mallow plants on which the caterpillars feed.

By early afternoon, we have had enough and after parting farewells, with the sun still high in the sky, I set off to cycle home. My journey to this wood ends with a super-fast whizz down a long straight hill, which means the return necessarily begins with a relentless uphill slog. Giving myself a needless challenge of getting to the top without stopping means that by the time I am on the flat with the luxury of freewheeling I can do little more than balance and breathe. As my heart rate returns to normal and I think about resuming a more purposeful pedal I happen to notice a few stunted common mallow plants growing in the roadside verge. There are only a few plants but I can't let the possibility pass. I need to return after dark to see if any nocturnal caterpillars were out having a nibble. At this time of year, this means waiting until at least 10 p.m.

I love the mystery and quietness of being outside at night, but prefer my wanderings to be away from roadside verges, with their passing traffic and possibility of people. So, very slightly on edge and keeping my bike close at hand, I begin to search the same patch of plants I had identified earlier. Almost immediately my torch beam picks out a slightly rough-textured green caterpillar, with pale lemon stripes down each side punctuated by small black dots. It seems almost too easy, for this is exactly what I am looking for. I take plenty of photos and race home barely noticing the hills or any passing traffic in my excitement.

A little after midnight, I send a photo to Roy Leverton, one of my trusted moth advisors, to see what he thinks, before slipping quietly into bed. But I can't sleep. My head is buzzing with possibilities. If this is a Mallow moth cater-pillar, there must surely be more of them. How widely spread are they? Where else does the foodplant common mallow grow locally? How could I find more? An hour later, still unable to sleep I check my emails. Roy, a self-proclaimed night owl, has already replied. 'How wonderful,' he enthused. 'This is indeed a Mallow.'

In the weeks that followed, whenever I could I pedalled out on my bike by day looking for common mallow plants in roadside verges. Returning to them after dark, I found more caterpillars. One night my husband accompanied me, our first night out together for many years. We counted twenty-five caterpillars along a 100-metre stretch of track, instantly elevating this verge into Scotland's premier location for Mallow moths.

Common mallow is a plant of waste land and margins. It likes a bit of disturbance to keep potentially smothering vegetation and scrub in check, but it also needs time for some uninterrupted growth to thrive. Across much of the UK, road verges get cut to within a centimetre of their lives at regular intervals throughout the spring and summer.

During my few weeks of searching, verge cutters were out in force. I'd return to a promising patch of plants after dark to find that in the intervening hours they had been scalped, with only those stems and leaves that hugged the ground closely being saved from the blades. One night I watched a Mallow moth caterpillar crawl up a severed stem, disappointingly devoid of leaves. It might have worked it out for itself, but nonetheless I repatriated it to the best-looking leaves I could find nearby. I hope it found enough leaf to fill its hunger and make it to a pupa. With luck, within the dark confines of this pupal case it would safely leave juvenile life behind and, through the miraculous transformation of metamorphosis, turn into a beautiful brown-medleyed adult.

CHAPTER 6

The Change

Omnia mutantur, nihil interit
(Everything changes, nothing is lost)

Metamorphoses, Ovid

Late in April, in the damp darkness of a culvert where over a hundred Herald (*Scoliopteryx libatrix*) have spent the winter, I find a pair of moths joined in a silent end-to-end embrace. I carefully coax them into a pot. Back home they are rehoused in a large cage made from an old chest of drawers. I add a potted willow sapling and position the cage in a quiet corner of the sitting room where I can keep an eye on proceedings. Herald is a moth that I have been getting to know well during the winter months, but I want to learn more about the life of this beautiful insect at other times of the year.

The next evening, after darkness has descended outside, the female moth wakes up and starts flitting around her cage. I pull up a chair nearby. A little later, as my household settles into sleep, I illuminate the arena with an unobtrusive dim red light. With the silence punctuated only by the ticking clock, I watch as she hovers with purpose around the plant, pressing the end of her abdomen against leaves and stems. In the morning, I count her legacy: just over a hundred small white eggs.

About a week later, the eggs have developed the palest cream tint and show within them a tiny brownish speck. Under a microscope this speck reveals itself as gaping jaws opening and shutting like a cartoon fish; behind them tiny black dots of the developing eyes are visible. Beyond this head end, a pale maggoty body coils, flexing occasionally. All these years obsessed with nature and I'd never thought to look at a butterfly or moth egg under a microscope. Even my teenage children are impressed.

The following day, just ten days after the eggs were laid, I watched, again under a microscope, the extraordinary moment a nascent caterpillar nibbled its way free from its egg.

Over the next twenty-four hours other eggs followed suit. Tiny translucent caterpillars were born. Some tucked into the leaf they found themselves on straight away, others chose immediate adventure and abseiled on a silken thread to a different leaf. Occasionally this led them adrift and I'd discover them searching in vain for something nutritious on the back of the nearby sofa. In the wild, silk threads are used like a kite, carrying the tiny caterpillar aloft in the breeze, hopefully to new foodplants. A risky dispersal strategy but if their foodplant is widespread, or the caterpillar not fussy, it can work well. Even so, many short lives must end through starvation when they fail to alight somewhere suitable.

Once leaf eating began, growth was rapid. The potted sapling was soon defoliated and I had to harvest a regular supply of willow leaves from outside. The tiny beings grew longer and greener and plumper, until about seven weeks into life and about five centimetres long, their behaviour changed. They became lethargic and sought refuge in the leaves at the bottom of their cage. A couple more days and the soft green caterpillars had transformed into hard, matt-black pupae.

For a month, the dark pupae, apparently lifeless, gave nothing away, despite my preoccupation with checking up on them. Then one afternoon I noticed one sporadically flexing in anticipation of something. Not wanting to miss the moment, I took it in its tissue-lined box with me as I went about my afternoon. Nothing much happened. I waited. Still nothing. Perhaps today wasn't the day. I was beginning to feel foolish for letting this small object take over my attentions and was on the verge of leaving it to its own devices when, with unannounced suddenness, although this was just the next moment in a long sequence of biochemical events, a small split appeared, revealing a sliver of russet-brown fur.

After a pause of a few seconds the split widened and a furry head was born, followed by six striped legs flailing in their unaccustomed freedom as they sought purchase on something solid. The new, or perhaps more accurately repurposed, life form managed to flip itself over and its legs made contact with the floor of the box. Shrugging off the pupal case, the crumpled moth hurried with inbuilt purpose, brown and orange button wings already lengthening, to find a vertical surface from which to hang and allow its wings to expand unhindered. This is a critical stage, a race against time for the moth, for if the wings are not given the space they need, they might dry in a stunted or twisted position and effective flight is never realised. Over the next half hour, the buds of wing slowly unfurled, the antennae flexed and the mouthparts gradually assembled into a tight coil. Before my eyes a perfect moth took form. Bearing no resemblance to the green leaf-devouring caterpillar it had been a month before, but with exactly the same DNA, it was finally ready to fly into adulthood.

Herald emerging

Moths, along with many other insects, undergo what is known as complete metamorphosis. The young stages, the larvae or caterpillars, are worm- or grub-like and very different in form from the winged adults they later become. This allows each stage to be adapted to specific functions without compromise or competition: the larvae to feed and grow and the adults to disperse and reproduce.

So different is a winged adult moth from its caterpillar that it isn't hard to understand why early naturalists considered them different creatures. Caterpillars seemed to appear on leaves from nowhere, and when they died, they left a pupa. If cut open, the pupa oozed goo, but if left alone, a winged butterfly or moth would later emerge. Here was a different animal, apparently reborn from a dead caterpillar. It was puzzling and hard to explain.

It wasn't until the rapid scientific and philosophical advances of the seventeenth century that the process of metamorphosis started to be unravelled properly. In the 1660s, Jan Goedaert, a naturalist and painter in the Netherlands, published three volumes depicting the transformations of a range of caterpillars which he observed from life. Although he was often unclear about how the caterpillars came into being, his illustrations clearly show that different types of caterpillars turned into different types of moths.

Towards the end of that decade, with smelly experiments involving rotting snake, fish and deer meat, Italian biologist Francesco Redi proved that maggots didn't spontaneously generate from the air. Instead, adult flies laid eggs which produced maggots and eventually pupae which turned into more egg-laying flies.

Also at this time, Dutchman Jan Swammerdam, peering down the newly invented microscope, skilfully dissected silk-moth caterpillars and pupae and found rudimentary parts of the adult moth inside the caterpillar. Convincing evidence that the caterpillar and eventual moth had to be the same insect.

Avidly taking all this in was a remarkable young German woman, Maria Sibylla Merian. Born in Frankfurt in 1647 to a family of artists and printmakers, she went on to become a talented painter herself,[1] but her real passion was natural history, and in particular plants and insects. As a young teenager Merian kept caterpillars, documenting the life cycles of species she found in her local countryside with detailed notes and watercolour illustrations. She observed that moths laid eggs that developed into caterpillars that turned into apparently lifeless pupae, out of which flew moths that mated and laid eggs.

In 1679 she published her first book on caterpillars. Her paintings, using living specimens, show the complete life cycle of various butterflies and moths along with the plants they fed on. Showing the real-life relationship between insects and plants was groundbreaking. The concept of ecology, the interactions between animals, plants and the environment now so fundamental to our understanding of the natural world, was barely considered at the time.

Maria Sibylla Merian's fascination with the life cycle of butterflies and moths stayed with her throughout her life.

She earned money as an illustrator and by teaching water-colour and embroidery to girls, but the study of insects occupied all her free time.

Towards the end of the seventeenth century, she and her two daughters moved to Amsterdam and opened up their own studio. They became part of a vibrant community of scientists, artists and thinkers; the city was a stimulating place to be for those with a curious mind. Ships arriving from around the world docked in the busy seafront and unloaded their cargo. Maria marvelled at the plants and animal specimens she saw, but became increasingly frustrated with their lack of context: insects displayed with no clue of the younger stages, or the plants or environment from which they came.

Spurred by a determination to discover more about these insects' lives for herself, at the age of 52 she went to South America, and what was then Dutch Suriname. In the hot, sticky heat she spent long days finding and sketching the plant and animal life she saw, always with a focus on insects. She reared the caterpillars she found, documenting the moths that emerged, despairing at the mould and ants that sometimes destroyed the pupa before it hatched. Notebook after notebook was filled with observations and paintings, and crate after crate with specimens. Annoyingly, after only two years and just as she was starting to make bolder explorations deeper into the jungle, illness forced her to return to Amsterdam. As she recovered her health, she organised her notes and specimens, finally creating the book for which she would later receive most acclaim. *Metamorphosis Insectorum*

Surinamensium was published in 1705 with sixty large copper-plate engravings illustrating the stages of development of many different insects arranged around the plants she had found them on.

Her reach went far. Her work was circulated, discussed and admired by the scientific elite of the Royal Society in London. Tsar Peter the Great acquired a large collection of her work. Later George III bought a first edition of her Surinam book for the Royal Collection. Carl Linnaeus used her illustrations to help him describe species of plants and animals. At least nine animals now bear her name.

Sadly, after her death some of her findings were disputed. Inaccurate copies of her books had been made and when these errors were spotted her work became widely criticised. Genuine observations such as a large spider capturing a bird were dismissed as fanciful female imagination. Only one hundred and fifty years later, when the explorer Henry Bates proved her bird-eating spider was accurate, was the record finally set straight; but her books and their legacy were soon forgotten.

In the 1970s Maria's work was rediscovered and she has since started to be justly remembered for the important contributions she made to science. Without doubt this amazingly driven woman advanced the understanding of metamorphosis and laid the foundation stones for what has become ecology. Her artistic talent is plain to see but she was also at the cutting edge of the natural science of her time.

★ ★ ★

My grandma was a kind supporter of my childhood natural history exploits. In her spotless house, polished glass-fronted bookcases were crammed with books she had read (Grandpa preferred the *Financial Times*). An old sketchbook gave a glimpse of her early married life when she explored the local countryside by bicycle, discovering the birds and flowers for herself. She was content and comfortable in her domestic role, but my parents always had the sense she was left slightly unfulfilled. In a different era, she might have been a successful academic.

Grandma had originally introduced my older sister to the Young Ornithologist's Club, then the junior club of the Royal Society for the Protection of Birds, but a few years later I was the one who took on the subscription. One birthday she bought me a nest box to put in my garden, and several years later a barn owl box which at the time I didn't quite know what to do with. Another birthday I was given a large plastic tank, a vivarium, which I variously used for tadpoles, toads, teddies and slugs. Once, as I sat beside her on the sofa, engulfed in her familiar scent of perfume mixed with lipstick and foundation, she drew a sketch of a woodland scene for me with a small girl sitting, her back against a tree. This was how I should watch wildlife in the woods, she said. If you sit quietly against a tree the animals won't know you are there. One year she gifted me two small, cotton wool-filled boxes, in each of which nestled a pair of moth pupae. Two of them were a smooth dark brown colour, the others were paler, coarsely haired cocoons.

In my hazy memory, they are no more nor less distinct

than other gifts I recall her giving me over the years, though I do remember feeling a small hint of responsibility when I learned what they were and duly kept them 'safe' on the top shelf in my bedroom. But such is the busy life of a ten-year-old that I wasn't very good at remembering to check up on them.

One day there was much excitement when I noticed a Buff-tip moth (*Phalera bucephala*) sitting on the ceiling; judging by the flimsy case left, clearly a product of one of the dark-brown pupae. The moth looked just like a piece of twig. I had never seen anything like it and after admiring and no doubt some petting of it, carefully carried it to the freedom of outdoors. Although I can still just about picture the wiry pale frizz that surrounded the other pupae I have no recollection of what, if anything, hatched out. Maybe the cat ate them. Looking back, it feels like a wasted moment of childhood. Although it makes me feel more relaxed about the raft of half-finished projects my own children have left in their wake, as an adult I'm now rather curious to know what my pupae became and wish that they'd left a more detailed impression.

Pupation and metamorphosis, as with so many big upheavals in life, are controlled by hormones. Throughout its life as a caterpillar, juvenile hormone rules. This keeps the caterpillar as a caterpillar. To start off pupation another hormone, ecdysone, is brought in to play. This hormone has already made brief appearances during the caterpillar's life to trigger the regular skin moults, but each time more juvenile hormone is released afterwards to maintain the caterpillar form. At pupation, juvenile hormone production is suppressed and ecdysone stays in control.

The first outward sign that a caterpillar is approaching pupation is a behaviour change. It stops eating and starts crawling around to search for a place to hide away. Each caterpillar species has its own preferred type of place it chooses to safely transition to adulthood.

This could be between leaves that it fastens together with silk, or buried in the soil, or hidden behind loose bark. For additional protection, some moth caterpillars spin a silken cocoon that provides an extra physical barrier to protect the developing moth as it pupates. This is something that butterfly caterpillars never do.

Safely hidden, the caterpillar body appears to shrink and it moults into a hard-skinned pupa (known as a chrysalis in butterflies). Most moth pupae are shaped a bit like a bullet with a blunt head end and more pointed tail. Already grooves and patterns on the surface hint at the design of insect that is being assembled within. Body segments are clearly demarcated, the shape of wing buds can be seen on the back and the outlines of fine, snaking antennae run down the front. Two circles where the eyes will be are separated by a bulge that will house the tongue. At the tail end there are tiny elegantly curled hooks called cremasters. These help hold the pupal case in place so the emerging moth can escape its confinement more easily.

Once a pupa, the moth can get metamorphosis fully under way. Certain cells start dividing rapidly to make wings, antennae, reproductive organs and other essential components of the adult body. Some caterpillar parts only require slight repurposing to better suit the adult's needs. Elsewhere, enzymes break down bits no longer required, leaving an amorphous soup of chemical building blocks from which new body parts are made. Muscles take form to power the wings, new neural pathways sprout, the digestive system once so efficient at processing plant tissue is reformed to deal with nectar. The actual time needed to effect the biophysical changes from caterpillar to moth is usually only a matter of weeks, but in some species of moth the pupal stage can last much longer; in some cases, several years. Surplus time is spent with a pause button pressed, waiting for a specific cue, usually a combination of day length and temperature, either to trigger the start of the cellular changes of metamorphosis or to trigger the final changes prior to emergence. This ensures that the adult will emerge at the right time of year and in synchrony with others of its species; essential for successful pairing and reproduction.

Moth pupae come in slightly different shapes, sizes and colours, but are hard to identify to species and usually so well hidden that they are difficult to find anyway. Early lepidopterists were unfazed by this, and without the convenience of modern light traps, seeking pupae was the most reliable way to encounter some species of moth. It also offered a way to get a pristine newly emerged specimen for the collection.

The activity, known as pupa-digging, requires taking a trowel to the base of walls, trees and other plants to carefully sift through the topsoil. Towards the end of the nineteenth century, Joseph Greene wrote a verbose guide to help would-be treasure seekers. Although he almost certainly exaggerates the success of some of his many expeditions, his closing words, when I finally get to them, do at least ring true: 'I conclude with one, literally one, word of advice to the incipient pupa digger and it is this: PATIENCE.'[2]

It seemed at least worth a try, so, one sunny December morning with nothing better to do, I decide to give it a go.

I choose the steeply sloping birch woodland where Mark Cubitt and I had enjoyed Scalloped Hook-tip in our light traps just six months earlier. Winter has rendered it a landscape of grey rock and grey lichen punctuated by hummocks of dark green moss. The twists and gnarls of the branches are accentuated without their leafy foil and the once verdant understorey is now mostly twiggy heather and shrivelled brown bracken fronds. I wonder how many offspring of our summer moths are slumbering beneath the barrenness, as caterpillars or pupae.

With the low sun on my back, I settle by a large birch tree and set to work with naïve optimism. I tease the soil away from fibrous roots, probe my way beneath thick mossy cushions and scrabble through loose gritty earth, my fingernails quickly becoming compacted with dark brown crescents. I feel like a clumsy squirrel looking for lost nuts, or perhaps a misguided pig in search of truffles. I wonder if you could train a dog to seek pupae?

I try different trees, different sides, various distances from the trunk, but I remain resolutely unsuccessful. At least it is enjoyable; a reason to be out, the mundane task therapeutic. I absorb both the warmth of the sun and the cold of the earth, listen to the thin tseeping of birds in the branches above and reassure myself that the reward, when it comes, will be all the greater for this effort. Even so, after three-quarters of an hour I'm starting to get a little impatient; it would be nice to find something.

I try a different tack at a large wizened oak. Maybe the rough, deeply fissured bark is hiding something. Carefully lifting some moss, my eye picks out an unusual bump in the bark beneath. At last! The stubby end of a dark pupa, wedged into a thin crevice. With extreme gentleness I tease it out and into a moss-lined pot. I'll nurture it in the cool of the garage over the winter and see what emerges next spring.

With satisfied relief, I call it a day. Finding a comfortable-looking nook between buttress roots, I sit back with the sun on my face to enjoy a flask of coffee. I'm not sure I'll be making a habit of this approach to moth-hunting. It's slow going and, however careful I am, disturbing soil, bark and moss is more destructive than the outcome really deserves.

The following April, I return to the wooded hillside. Some time the previous night my pupa successfully became an adult Scalloped Hazel (*Odontopera bidentata*). This is an appro-priately named moth, with its triangular hazel-coloured wings sporting gentle scallops along their outside edge; effective leaf litter camouflage. Although it is a widespread and common moth which could probably be released happily in any of my local wooded places, I want to return it to the place of its youth. As I lift the lid off the pot it flutters weakly up and away, into obscurity among the crooked branches, newly decorated with fresh green unfurling leaves. Hopefully tonight it will enjoy success in the number-one task of newly emerged moths everywhere: that of finding a mate.

CHAPTER 7

Amorous Aromas

Coming from every direction and apprised I know not how, here are forty lovers eager to pay their respects to the marriageable bride born that morning amid the mysteries of my study.

'The Great Peacock' from *The Life of the Caterpillar*,
Jean-Henri Fabre

It is a chilly but sunny morning in early May. I'm trying to concentrate at my desk but it's no good. Just outside, my husband is noisily grinding metal on the driveway. Even more annoying, I must contain my irritation as he is admirably trying to repair the car's rusting bodywork in an effort to prolong its useful service to us. In the room next to me a daughter is having a piano lesson via Zoom, involving lots of loud and repetitive scales. What is it about scales, that they can never be played quietly? As grumbling frustrations froth in my head, I know the solution is to go and absent myself for a while to the fresh air. Although nearly midday, the temperature has only just passed 8°C, so I doubt there will be much in the way of insect life for me to find, but I take a net and some pots anyway and cycle towards the coast. I choose the direction the quiet farm tracks take me, with no particular destination in mind.

After fifteen minutes I'm at a tiny reed bed. Despite it being within a few kilometres of home, I only found this oasis last year; part of my Covid-19 lockdown-enforced explorations. Phragmites reeds flank a small stream for about a hundred metres as it flows through intensive arable farmland that I might otherwise avoid. Neat rows of cabbage seedlings stripe the length of one field; another has been ploughed to a fine tilth and arranged in steep ridges and furrows, perhaps already seeded with carrots. Along the stream though, the vegetation is left to do its own thing. Among the reeds there are old willow trees with deeply fissured trunks leaning towards the water, their smooth bendy branches reaching skywards with fresh green leaves unfurling. There are grasses and docks and lots of nettles, with stinging hairs poised to find any gaps between sock and trouser leg.

I prop my bike against a tree and make my way to the stream's edge. It's small and sluggish; more ditch than stream. It starts a few kilometres away, journeying slowly along field margins before slipping quietly into the sea just a few hundred metres from where I now stand. There is little sign of aquatic life as I peer into its stagnant shallows. Ahead some flapping and splashing is going on, so I crouch quietly to peer through the reeds. A splendid mallard drake warily looks back at me. Perhaps his mate is sitting on eggs nearby.

Not wanting to intrude, I start to retreat, but as I turn, I notice something that makes me stop and instantly forget about ducks. Low down on the trunk of the willow next to me are some small boreholes about the diameter of a pencil. For years I have been looking for holes like this, poking about in vain at the base of willow trees I happen to find myself next to. I'd even searched some of the willows near here last summer, but without reward.

These holes are the characteristic work of the caterpillars of the Lunar Hornet Moth (*Sesia bembeciformis*), one of our elusive clearwing moths.

Clearwings are a family of moths which, as their name suggests, have see-through wings. Adults are short-lived, mostly small, fly rapidly in sunshine and can't be enticed by light. Resting on a tree or sitting atop a flower, they look just like a fly or wasp and are easy to mistake for one, unless you know what you are looking for. The caterpillars are even harder to see. These pale grub-like creatures spend their lives out of sight deep within the stems or trunks of their foodplant where they feed on pith and woody tissues. They only see the light of day if their accommodation is felled, or when, as newly transformed adults, they crawl outside.

I examine more trees and discover more holes. Then I notice a tree with small mounds of orange-brown wood shavings beneath the holes. These are piles of caterpillar frass (droppings), expelled from their boreholes. It suggests recent activity; a tantalising indication that caterpillars or pupae lie hidden just a few centimetres away. Frustratingly the adults don't emerge until July. I'll have to wait a few more months before it is worthwhile staking out the trees for a Lunar Hornet Moth vigil.

Whether or not I discovered an occupied tree, my chances of seeing Lunar Hornet Moth had become much higher the previous year, when a chemical lure designed to attract males of this species became available to buy. The number of records of Lunar Hornet Moth in the UK soared as people tried it out. I watched vicariously as photos were posted and joyous moments of discovery described. Perversely I was reluctant to join the throng and jump on this particular bandwagon; it seemed too easy, a cheat's way to see a moth. Though of course, had I been invited, I knew I would have leapt at the chance to join somebody using a lure.

Moths do much of their communicating by scent, emitting volatile chemicals known as pheromones from special glands to signal information to other moths of the same species. The Lunar Hornet Moth lure works by imitating the female moth's sex pheromone, her chemical call for a mate.

In most moth species it is the female that releases a sex pheromone, a behaviour referred to as 'calling', when she is ready to mate. Her seductive scent wafts through the air and is picked up by an amorous male's antennae. Males are usually better endowed than females in the antennal department, some species with magnificently feathery plumes sprouting from their heads. Such finery gives a greater surface area of antennae and therefore more space for special scent receptors, tuned to pick up the appropriate female's pheromone. The quicker he can pick up the smell of a calling female, the sooner he can try to reach her. For the best chance of mating, he needs to be the first suitor on the scene. As soon as he detects the perfume of a potential mate, he flies towards the scent, taking a zigzagging approach to locate the strongest smell and home in on the source. This attraction works over surprising distances. Wind strength and direction must help, but there are reports of some species attracting a suitor from over ten kilometres away.[1]

Lunar Hornet Moth

Once he has arrived at the female, other chemical signals are exchanged between prospective partners and maybe some acoustic or visual communication too. These allow any last-minute changes of heart before mating takes place. Early lepidopterists were well aware that a female moth did something to attract males to her. Before they fully understood what was going on they used the behaviour to good effect for themselves, as a way of getting male moths for their collections; a technique known as assembling.

The collector would enclose a newly emerged female moth in a mesh-sided cage, and take her to a suitable site. As she released her pheromones any male moths in the vicinity would come flocking. With only one thing on their mind, the males are not easily distracted, and the waiting entomologist could collect specimens of choice as they jostled around the caged female.

Towards the end of the nineteenth century, French entomologist Jean-Henri Fabre had a go at trying to work out how the attraction worked. He narrates his efforts in one of his superb natural history books, *The Life of the Caterpillar*.[2] His investigations began on a summer evening with urgent summons from his young son Paul, to come up to his bedroom to see a room-full of moths, apparently as 'big as birds!' That morning, a female Great Peacock Moth (*Saturnia pyri*) that Fabre had reared from a caterpillar had hatched out in his study. As the chaos settled, it dawned on him that this female was likely responsible for the winged invasion of his son's bedroom.

As their name suggests, Great Peacock Moths are large. With a wingspan of up to twenty centimetres they are the largest moths in Europe. They have a striped fluffy body and large smoky-grey wings edged with a bold creamy-white border and embellished by four striking eyespots. The sound and the sight as they fluttered into Fabre's house from the dusk must have been memorable; apparently the flustered maid at first thought they were bats. Having established that it was indeed the female that was the cause of the moth mayhem, Fabre spent the following evenings, and indeed consecutive nights in following years, trying to work out how she managed so successfully to attract the males to her side. Through experiments variously involving cutting off the male's antennae, shaving scales from their bodies and encasing females in boxes or bell jars, he eventually concluded an 'odour' from the female rather than sight or sound was most likely responsible, yet he never got to understand fully what was going on.

Further speculations came several decades later, towards the end of the nineteenth century when New York entomologist Joseph Lintner put a female Spicebush Silkmoth (*Callosamia promethea*) on the window ledge just outside his office. It wasn't long before there were tens of male moths clamouring around her, as well as a congregation of people watching in awe from the pavement below. Lintner thought the attraction was probably derived from a chemical released by the female that the males detected with their large feathery antennae. He went on to muse that this chemical could become a useful tool for pest control, if only chemists could synthesise it.

Lintner didn't live to see it, but such a chemical was finally characterised in 1959; the sex pheromone of the domesticated silkworm. The research was an epic undertaking led by

German biochemist Adolf Butenandt (who had previously received the Nobel Prize for discovering sex hormones in humans). He and his colleagues spent twenty determined years and used half a million female moth abdomens to isolate and work out the structure of the chemical. They named it Bombykol. In subsequent decades, as equipment in biochemistry laboratories has improved, the process has become easier and quicker. We now know the chemical makeup of pheromones from a huge number of moth species and can synthesise our own effective imitations of many.

The primary driving force (for which you could read funding) for scientists to produce artificial sex pheromones has been for their use in pest control, just as Lintner once reflected. Caterpillars can wreak havoc on commercial crops in agricultural and forestry yet using pesticides is costly, non-discriminatory and leaves toxic chemicals in the environment. Artificial pheromone lures provide an alternative. The attracted males can be trapped and killed, hopefully before having a chance to mate. The lures can also be used to monitor how many moths are around, enabling growers to limit pesticide applications to when they are really needed.

The process of making artificial pheromone lures is, in theory, quite straightforward. First, pheromone glands are dissected from the body of a few female moths and mixed with a solvent to dissolve all the pheromone chemicals. The liquid is then passed through sophisticated chromatography equipment, which separates out its different chemical constituents.

The next stage is to work out which of these different chemical components are the best at getting a male moth excited. This is where an electroantennogram, or EAG, comes in. An antenna is cut off a male moth and wired into an electrical circuit. Each chemical is wafted in turn over this excised antenna and if the receptors on it recognise the chemical, they get excited and produce a tiny electrical current. Normally this electric pulse would be sent to the moth's brain to be interpreted and acted upon but in an EAG the tiny signal is amplified and produces a sudden spike on a machine readout. Bigger spikes suggest a stronger response of the moth to that chemical.

Once the identity of the chemical or chemicals which most excite the antennae have been identified, a mix is synthesised using off-the-shelf reagents and the cocktail tested on living moths. If it works, the mixture is made into lures for finding male moths in the wild.

This technology has been used widely to help control pest species, but it is now used increasingly by conservationists too, to help save moths. Pheromone lures are effective even when moths are at very low density which makes them ideal for surveying rarities. Unlike cumbersome light traps with heavy batteries or generators to power them, lures are small, lightweight and can be deployed easily, even in remote areas. Captured moths are quickly released unharmed and artificial lures, as long as they are used carefully and sparingly, have no lasting impact on the moth's subsequent behaviour or reproductive success.[3]

★　★　★

In Britain, artificial pheromones are being used to support conservation of two endangered moths, the Kentish Glory (*Endromis versicolora*) and Barred Tooth-striped (*Trichopteryx polycommata*). Researchers from Canterbury Christ Church University in Kent, the University of Greenwich and Butterfly Conservation, with help from scientists in Sweden, have been working together to develop effective lures for these species.

Joe Burman is leading the lures' development. He told me the process had been unusually easy for Barred Tooth-striped, taking just two or three years from identification of the pheromone to implementation of an effective lure. Apparently, in the pheromone lure-making world, that is quick. These lures have since been used to identify new sites for this moth and have made monitoring their numbers and understanding more of their ecology much easier. On the other hand, for Kentish Glory, getting a good result is proving more difficult.

Kentish Glory is a moth that has a name and appearance to match. It is a large moth, by British standards at least. The female's chunky body is about three centimetres long, tented by striking brown and white wings. The male is a little smaller with impressively feathery antennae and more of a brick-orange coloration. On warm sunny days in early spring, he zips optimistically around young birch trees in search of females. Although once thinly scattered across the southern half of the UK, the moth is long gone from England and Wales. Now the only place it lives in Great Britain are small pockets of the Scottish Highlands.

Even more worryingly, despite increased efforts to look for it, it seems to be getting rarer. Nobody is quite sure why.

A hundred years ago collectors may have had an impact. Its appearance makes it a covetable moth, and with males easy to lure by assembling, many were taken. Climate change could be another factor, but another nail in its coffin is its requirement for very specific habitat, which is disappearing in many places. In Scotland, Kentish Glory thrives in moorland areas with a good supply of young birch trees. After about ten years of growth, for some reason the trees are no longer suitable and the moths ignore them. Meanwhile, unless most grazing and browsing animals are excluded, new birch saplings find it hard to establish, which means no tasty leaves for the Kentish Glory caterpillars to feed on.

Each spring a team of volunteers heads out into areas where Kentish Glory is known or where the habitat looks right. They use light traps for adults at night and search suitable birch trees by day for eggs and later caterpillars. An effective pheromone lure would make surveying so much easier, and open up opportunities for more reliable long-term monitoring.

Although Joe Burman and his team have identified a pheromone mix which gives a good EAG response from Kentish Glory antennae wired up in the lab, in the field it has limited effectiveness. Males are attracted towards the lure, but they don't always come very close or land on it for long. There seems to be something missing. Wondering if male moths also use visual cues, one year some fake females were crafted from felt and pipe cleaners. They look remarkably realistic to me, but the male moths weren't fooled.

'We are still looking for the key to make this one work well in the field,' Joe confessed. It is a slow process as each year they only have a short window of about three weeks when the adult moths are on the wing, to try out a new tweak to the lure.

I live several hours' drive from the Kentish Glory's realm and have yet to orchestrate a trip which finds me in the right place at the right time to try to track down one of these magnificent moths. For the moment, it remains an experience to look forward to. Meanwhile, there are plenty of charismatic moths to entertain me just a short distance from my doorstep. In late June I return to the boreholes in my local willows. My husband and I cycle there during a work-from-home lunch hour. When we arrive, my heart sinks. The place is almost unrecognisable from my late winter visit.

Between the grassy path and the trees I want to look at, the once ankle-biting nettles have become a treacherous chest-high jungle. I battle my way through, soon giving up any hope of avoiding the stings which get me through my trousers, on the back of my hands and even reach my face. My husband looks on, I suspect more in bemusement than awe. The nettles continue their luxuriant growth around the base of the willow trees, smothering the bark from easy view, though maybe fortuitously also providing an effective barrier to hungry woodpeckers, which are known to enjoy feasting on Lunar Hornet Moth caterpillars and pupae living inside trunks.

It is too long ago for me to remember exactly which tree had the most promising-looking moth holes. Using a broken stick, I part some of the nettles from a nearby trunk but can't see very much; certainly no waspish moth waiting for me. I uncover some holes in the next trunk, but no obvious sign of any recent mothy activity. Admitting early defeat, I battle my way back to the path.

Luckily, I have a plan B. Thanks to this year's birthday gift, I am now the owner of a Lunar Hornet Moth lure. I uncap the small vial and suspend it above the nettles, from the branch of one of the willow trees.

Within five minutes I am watching my very first Lunar Hornet Moth. It happened much more quickly than I had imagined, and required no skill beyond identifying likely habitat, but it remains a memorable moth-ing moment in my life. I'm told I even did a little jumping for joy; I must have forgotten myself for a moment. Even though I knew what I was waiting for, at first sight the insect that arrives to fuss around the lure convincingly fools me into thinking it is a large yellow and black wasp. I instinctively take a step backwards. Not only is the colour and shape right, it does a reasonable imitation of a darting, hovering vespid flight. We watch in amazement, marvelling at how a moth can evolve to look so waspy, and how a minute amount of invisible chemical can be so effective at summoning one. My husband, admittedly no entomologist, finds it hard to accept that this insect is indeed a moth. I have to agree, there is little to suggest moth in the insect we see before us. Only when I coax one onto my finger and it pauses long enough for me to take in the furriness of its legs and body, do I notice its wings and antennae are not truly wasp-like either.

After a few minutes I get uncomfortable watching them getting confused by the promise of a female they can't find.

Pheromone lures have revolutionised our understanding of where clearwing moths occur and provide people with unrivalled experiences seeing these wonderful insects up close. But just as with any form of moth recording, from taking photographs to using a light trap, lures need to be used sensibly and thoughtfully. Butterfly Conservation has published a code of conduct, including guidance on when and for how long a lure should be used in one place. Moth welfare should always be put above moth recorders' desire.[4]

I doubt anyone will be coming to these willows any time soon, and it's even less likely that they will come wafting moth pheromone lures, so these clearwings shouldn't be side-tracked by any more fake females in their short lives. However, I decide that they have enamoured me for long enough. We pack up, leaving the boys to devote their efforts to pursuing the real thing. Though for all I know, they've already had a successful morning doing just that.

CHAPTER 8

Sweet Tastes

Their probosces were deep plunged, and as they drew in the
sweetness, their wings quivered slightly as if in ecstasy.

'Reading' from Collected Essays, Virginia Woolf

As night falls, the garden or back doorstep has much to recommend it. When the children were younger it was a place to unravel after the crescendo of noise and chaos that surrounded bedtime; to briefly shut the door on things that need to be done. I vaguely remember my parents disappearing into the dusk together to 'check on the vegetable patch' at the end of the garden. I wonder now if the vegetables might have been a foil. Maybe they simply needed somewhere to go.

My own garden is far smaller than that of my childhood, but it's still a place of evening calm despite long giving up on bedtime battles. On mild, still nights, I patrol the flower bed, shining a dim torch over plants and the awakening night-life. With luck, flowers which not many hours earlier hosted hoverflies, bees and wasps will be entertaining a clientele of moths with their sweet cocktails.

Some hover just above the flower, wings a-blur, drinking briefly but forever moving on. More engaging are the ones that settle themselves on a flower head, front legs braced, pollen grains dusting their furry bodies while the proboscis elbows and flexes to seek its reward. The dexterity required to manoeuvre this spindly tube, frequently as long as the moth itself, to the required position is impressive.

The very first moths, flying around 200 million years ago, had chewing mouthparts and probably fed on fern spores and pollen from primitive conifers in their prehistoric swampland homes. To keep hydrated they might have sipped on dew and other small liquid droplets and it is thought this gradually led to the development of more specialised sucking mouthparts to better deal with these food sources. Once flowering plants made an appearance, many millions of years later, things started to change more rapidly; at least on evolutionary timescales. As a smorgasbord of new dining opportunities arose, in particular through the provision of nectar, a more suitable feeding apparatus was needed to access it.

There are still tiny moths that eat fern spores and pollen grains, using a special cavity in the mouth to process these granular foodstuffs. But most others have moved on, with the evolution of a long tubular mouthpart called the proboscis; a structure a bit like a drinking straw but with additional sponge-like properties. Importantly it can reach deep into flowers to access the pools of nectar there. It is made of two halves, each c-shaped in cross section, which zip together soon after emergence from the pupa to form a sort of tube. At rest the proboscis is typically tidied away in a tight hose-pipe coil under the head, only being unravelled for feeding.

To make seeds to reproduce, plants need to get pollen grains from one flower into another of the same species so they can fertilise the ovule. Some take their chances with the wind but the vast majority enlist animals to carry pollen between flowers. This is where nectar comes in.

Nectar is produced to provide an enticing reward, encouraging the flower visitor to climb into the flower to access the sweet drink, where it also more easily picks up or deposits the sticky pollen grains. Many animals are effective pollinators, but insects are the most abundant. Bees, flies, wasps, beetles, butterflies, moths and ants are among the six-legged creatures in on the act, and, with their slightly different modes of feeding and different plant preferences, each has an important role to play.

Although moths have long been recognised as pollinators, just how vital their role is in helping plants reproduce is only beginning to be realised, as scientists explore insect–plant interactions taking place under the cover of darkness in more detail. Moths' long mouthparts allow them exclusive access to nectar from flowers that other insects can't reach and their furry bodies pick up pollen grains easily. Many species are good fliers with the potential to carry pollen relatively long distances, and those working the night shift provide a continuation of service while daytime pollinators roost.

★   ★   ★

The lengthiest proboscides of the moth world are found in some of the hawk-moths, a large family of impressive-looking insects, with thick bodies, large eyes and powerful wings. There are over 1,400 species of hawk-moth worldwide. Not all of them feed as adults, but almost all that do are furnished with a proboscis of impressive length, allowing them to reach deep into flowers to access nectar rewards and in the process pick up pollen.

One of the most impressively endowed, with a proboscis nearly thirty centimetres long, is an African hawk-moth (*Xanthopan praedicta*), which was predicted to exist long before it was actually discovered.

The story involves two extraordinary naturalists: Alfred Russel Wallace and Charles Darwin. Both men spent time in different parts of the tropics in the middle of the nine teenth century, observing and documenting the enormous variety of plant and animal life that they encountered and asking questions about how such diversity could have come to exist. Charles Darwin is now one of the world's best-known biologists, credited with coming up with the theory of evolution by natural selection. This theory, which explains how different species have evolved over long periods of time from common ancestors, has become central to our understanding of life on earth.

Natural selection favours individuals with characteristics, or traits, that help them to survive and reproduce, and so it is these traits that more likely get passed on to the next generation and come to persist in the population. This provides a mechanism for populations of animals and plants to gradually change their looks or behaviours over time to become adapted to new environments, ultimately giving rise to new species.

When Darwin was contemplating his theory, ideas of the Western world were dominated by the Church, and the widely held belief was that all life was made at once by a divine creator. Even though it was becoming increasingly evident that this creationist theory had flaws, it remained heresy to declare otherwise.

Scared of damaging his reputation by suggesting such a controversial alternative, Darwin sat on his uncomfortable ideas for twenty years, hoping to gather more evidence before speaking out. Meanwhile, thousands of kilometres away in a hot Malaysian jungle, Wallace was also mulling over natural diversity and independently came to the same conclusions. In 1858 he wrote to Darwin to excitedly share his ideas and seek advice for publishing them. This proved the nudge Darwin needed to go public himself, and later that year the two men made a joint announcement of their thoughts and evidence at Britain's leading natural history society, the Linnean Society of London. The following year Darwin published his important book that had been on the back burner for years, *On the Origin of Species*. The rest, as they say, is history.

One of Darwin's many natural history interests was orchids. In 1862 he was sent a star orchid, *Angraecum sesquipedale*, from Madagascar. It has a pale moon-white flower behind which stretches a long narrow spur where the nectar collects. It was this long spur, about thirty centimetres deep,

that caught Darwin's attention. He knew orchids were pollinated by insects, but puzzled over what species could possibly
reach this nectar reward and help pollinate the flower,
famously exclaiming almost as soon as he had set eyes on
it, 'Good Heavens, what insect can suck it?' He decided that
it had to be a moth with a very long tongue. Wallace
supported Darwin's suggestion but went further, noting that
there was a hawk-moth, *Xanthopan morganii*, which was found
in east Africa and had a suitably long mouthpart. He
published a paper in which he said, 'That such a moth exists
in Madagascar may be safely predicted; and naturalists
visiting the island should search for it with as much confidence as astronomers searched for the planet Neptune – and
they will be equally successful!' He even provided an illustration of the predicted moth.

It took over forty years, but both men were eventually
proved right. In 1903 a closely related hawk-moth was discovered on Madagascar, although it wasn't until nearly a century
later that a team of scientists provided photographic and
video proof showing that this moth actually fed from and
pollinated the orchid. Nowadays the orchid and moth are
sometimes known as Darwin's Orchid and Wallace's Sphinx
Moth; an equitable tribute to these influential scientists.[1]

Large hawk-moths appear in Britain too. With a wingspan
of around ten centimetres the Convolvulus Hawk-moth
(*Agrius convolvuli*) is not quite the dimensions of its exclusively
tropical cousins, but impressive nonetheless. The first time
I saw one I thought it was a large mouse. Stumbling to my

garden light trap in the dewy half-light of dawn, bare-footed, in pyjamas and with barely awake eyes, I noticed a greyish lump resting on the side of the trap. It didn't scuttle away and I suddenly realised what it was: a very large moth. With a jolt of adrenaline, I was at once very awake.

It is a magnificent creature. The long narrow wings are dusty grey, streaked with daggers of black which give it a sleek yet subdued elegance. Beneath these wings it hides a chunky cigar abdomen with pale pink, white and black stripes down each side. To contain it safely so I can parade it in front of my children later, my usual pots are of no use. Rushing back to the kitchen, I rummage through the drawer of empty margarine tubs for something more suitable. As I carefully lift it into the pot it raises its wings a little. Where you might expect each armpit to be, a bold fuchsia-pink patch is revealed, looking a bit like a pair of glowing devil eyes.

Convolvulus Hawk-moth is one of the largest moths found in the UK. It isn't a permanent resident as it is unable to survive the chill of our winter. Even so, several hundred are usually recorded here each year, mostly in late summer, having migrated in a broadly northwest direction across Europe. Each is met with great excitement by its finder. Frequently they are seen near light traps rather than in them, perhaps resting on garden furniture or walls. Intriguingly they often land on clothes left hanging on a washing line which has led some hopeful recorders to leave freshly laundered towels out by way of a fragrant lure. Plenty of discussion follows, over which washing powder or fabric softener brands might have the best hawk-moth attracting power. As far as I know, no conclusions have yet been reached but I sometimes leave a tea-towel or two hanging out near my light traps in August and September, just in case.

About five years ago, the second Convolvulus Hawk-moth sighting of my life saved an otherwise disastrous day. In a rare spur-of-the-moment decision to do something as a family we had decided to spend a night camping in the Lammermuir Hills. I knew a good spot, just off a quiet road and with a large flat area that could easily accommodate our tents. It was a peaceful place that I had visited previously with moth traps, though this trip was to be an evening spent together rather than me disappearing off into the gloam. What I didn't anticipate was that on a warm summer's weekend other people would have a similar idea. We arrived to half a dozen parked cars and an assortment of tents. One extended family group was even kitted out with striped wind breakers, tables and chairs, blaring music players and yapping dogs running loose. There was plenty of space for us but it wasn't the quiet setting we were hoping for. In our typically indecisive way, my husband and I couldn't decide what to do. Surely they wouldn't be noisy all night? Maybe we could pitch our tents and go for a walk to seek solitude. Did either of us have the courage to ask them to be quieter? One of our children wanted to go home, the others wanted to stay. My husband showed little sign of being able to tolerate the noise for long and I agreed it was hardly relaxing. But mostly I just wanted everyone to be happy. Then our dog started barking at the other dogs.

In the end, after increasingly fraught discussion of 'options' we decided to go for a walk and return home to our beds to sleep. I can no longer remember the walk but the journey home was downbeat and quiet, with the evening's expectations dashed. But as we reversed into our driveway and the car's headlights swept over a large honeysuckle bush, I saw the unmistakeable bulk of a large moth hovering over the flowers. 'Look!' I squealed. Everyone did. We tumbled out, eager to get a closer look, and there it was: an enormous grey moth drinking from the honeysuckle flowers. Aware of the sudden audience, or already satiated with nectar, it didn't stay long before rising higher into the air and zooming out of sight down the hill. The night ended on a high and it is a moth encounter we all still remember.

Roy Leverton is well known in the Scottish moth community for his knowledge, opinions and provocative scepticism of anything he reads. With over six decades spent carefully observing moths in various parts of the UK, few can rival the length of his personal experience. Nowadays his efforts are mostly restricted to his small croft north of Aberdeen, but he maintains a talent for finding moths. Maybe, as another eminent moth-er once remarked, this could be because Roy had actually been a moth in a former life.

In my own early days of moth recording in Scotland, when puzzling over the identity of a worn indeterminate grey-brown moth, I was frequently told 'just ask Roy'. When I eventually plucked up courage to join the moth forum of which he is a pivotal part, to post my queries, his responses

Green Longhorn (*Adela reaumurella*)

Elephant Hawk-moth (*Deilephila elpenor*)

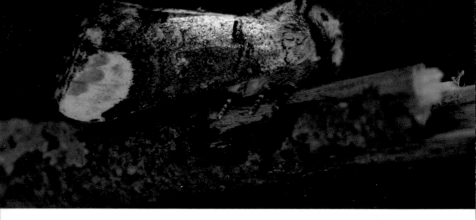

Buff-tip (*Phalera bucephala*)

Poplar Hawk-moth (*Laothoe populi*)

Buff Arches (*Habrosyne pyritoides*)

Herald
(*Scoliopteryx libatrix*)

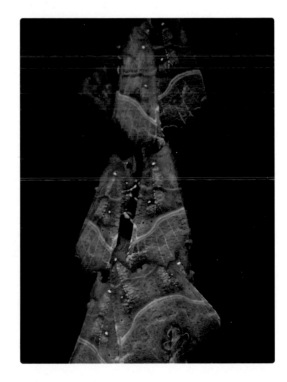

Merveille du Jour (*Griposia aprilina*)

Convolvulus
Hawk-moth
(*Agrius convolvuli*)

Peppered Moth (*Biston betularia*)

Six-spot Burnet
(*Zygaena filipendulae*)

Mallow moth caterpillar (*Larentia clavaria*)

Feathered Thorn
(*Colotois pennaria*)

Garden Tiger (*Arctia caja*)

Pimpinel Pug (*Eupithecia pimpinellata*)

Scalloped Hook-tip (*Falcaria lacertinaria*)

Canary Shouldered Thorn (*Ennomos alniaria*)

were rapid and helpful. He has patiently advised and humoured me in my mothy endeavours ever since.

Roy tells me his moth obsession began in his boyhood town of Salford. He describes Salford in the 1950s as 'more dark Satanic mills than green and pleasant land', yet despite thick pollution from coal fires and heavy industry leaving dirty sooty deposits everywhere, he still found moths. He noticed that some moths seemed to thrive in this environment and that those on blackened surfaces tended to have darker wings, making them better camouflaged. The more moths he found, the greater his interest became. Realising that some of his own observations were not in the books he read, he saw opportunities to make scientific contributions too. He was hooked.

Roy is unusual among modern-day moth-ers in that he devotes at least as much time, if not more, to finding his moths without a light trap. Growing suitable plants in his garden, to feed hungry caterpillars and fuel thirsty moths, is part of this. Each year, with a neighbour's help, he grows ornamental tobacco plants *Nicotiana alata* 'Sensation Mixed' from seed. A patch of his garden is devoted to the plants, for the single purpose of attracting Convolvulus Hawk-moth; they find the flowers' heady fragrance hard to ignore.

As soon as Roy hears that one of these moths has been spotted elsewhere in the country, he begins a nightly vigil of the patch. This doesn't mean a brief glance out of the window after dinner if he remembers. Instead, every night for about six weeks he's outside on patrol at dusk, prowling around for at least half an hour. Only high winds or lashing rain stop him. If conditions feel good, he returns several times through the evening and into the small hours of the following morning. Perhaps it's no wonder he's been lucky to see these impressive hawk-moths most years; you make your own luck with moths.

Roy also nurtures a swathe of campion. These are favourite flowers for many thirsty moths, but none more colourful than a duo of spectacular hawk-moths – Elephant Hawk-moth (*Deilephila elpenor*) and Small Elephant Hawk-moth (*Deilephila porcellus*). In straw polls on the nation's favourite moth, Elephant Hawk-moth frequently comes out top.

In Britain we are used to looking on enviously at tropical insects and birds, many of which seem to outshine our own residents in both size and colour. Warmer temperatures in the tropics mean cold-blooded insects can function with a larger body than in the cooler latitudes, but are tropical species more colourful too? As I look through a book of British moths there are plenty of colourful pictures, all sorts of impressive-looking creatures to tempt me, and none more so than in the hawk-moth section. Is it simply that familiarity dampens any wow factor, or a case of the grass is always greener and the moths far more colourful elsewhere? Of course, scientists have approached the question more objectively and report no evidence that tropical animals are more colourful than those from the higher latitudes. Biodiversity increases as you get closer to the Equator, so there is more variety of everything, but apparently the proportion of the dazzling to the drab is, overall, more or less same wherever in the world you are.[2]

Our two Elephant Hawk-moths are certainly proof that eye-catching colour exists outside of the tropics. They are clothed in a clashing punk-pink and camo-green combination, with natty white legs and large olive eyes. They are similar to each other in behaviour and looks but in Britain the Small Elephant Hawk-moth is generally less common and less likely to be found in gardens than is its larger cousin.

Roy well remembers his first Small Elephant Hawk-moth encounter in rural Sussex about fifty years ago. 'It was a magical evening. In the valley bottom was a patch of red campion next to some scrub, and five *porcellus* were feeding from flowers in the early dusk. It was light enough to watch them as they hovered. As a species I'd long desired, what better way to see it for the first time!' he reminisces wistfully.

Although they are not as common in Scotland compared with southern England, after moving north Roy only had to wait six years before adding Small Elephant Hawk-moth to his garden list. There, on 22 June 1996, he was delighted by one nectaring at his home-grown red campion in the early dusk. His strategic gardening efforts had paid off. Since then, he has continued to watch both species of elephant hawk-moths feeding at his flowers in the mild dusks of summer, marvelling at their aerobatics and forever aspiring to take an even better photograph.

Long tongues and a sweet tooth are common traits of most hawk-moths, but there are exceptions. Some species, for example Poplar Hawk-moth (*Laothoe populi*) and Lime Hawk-moth (*Mimas tiliae*), forgo feeding completely as adults, relying solely on reserves they acquired as a caterpillar to power their large bodies through their short reproductive lives.

Three species are furnished with decidedly short proboscides; only about a centimetre in length. They show little interest in flowers, instead meeting the needs of a sweet tooth by feeding on honey. These are the Death's-head Hawkmoths, large moths with a sinister reputation in folklore and literature across the world. Two were found in the bedchamber of King George III; blamed for throwing him into a subsequent bout of madness, though nobody really knows if the king actually saw the moths. Conspiracy theorists claimed that they had been put there on purpose. Dracula sends them to his thrall, R.M. Renfield, to keep him stocked up on snacks. John Keats mentions the 'death-moth' in his 'Ode to Melancholy', and in The Return of the Native Thomas Hardy uses the moth as a prophecy of doom. More recently pupae of the Death's-head Hawk-moth starred in the blockbuster horror film Silence of the Lambs, the accompanying poster affording the moth overnight fame.

All three species of Death's-head Hawk-moth have similar lifestyles. None is resident in Britain, but each year small numbers of one species, Acherontia atropos, turn up, having dispersed from the warmer countries of southern Europe, Africa and the Middle East. With a wingspan of 12–13 centimetres it is the largest moth to be found in the UK.

Death's-head Hawk-moths get their name from the pale skull-like marking on the back of the thorax. Not everyone is convinced it is a skull. Some see the pattern as a yellow

Darth Vader staring out at them, others as a monkey. For some reason I'm reminded of a golden retriever with doleful eyes. Of course, wearing a thumbnail-sized image of a skull on the back of its head is unlikely to impart much benefit to the moth. With a bit of imagination and viewed from the right angle, the markings are also slightly reminiscent of a bee. Could this 'death's-head' pattern have evolved to improve its disguise in the hive?

To get access to the sticky combs that they desire, Death's-head Hawk-moths must first get past and avoid the resident and protective bees. The moth's thick cuticle is relatively resistant to bee stings, but they also emit pheromones which make them smell like a bee, helping them to crawl into the hive unnoticed. Another unusual feature of these moths is their alarming squeak, made by inhaling and then expelling air over a flap between the mouth and throat. The purpose of the squeak is uncertain, but it is often made when the moth is threatened so it likely serves as a startling defence. The moths also squeak as they enter the hives and some suggest they could be mimicking the buzz of a queen bee, thereby furthering their disguise. Once they have successfully infiltrated, they use their short stout tongues to feed on the sweet liquid honey.

My first encounter with a Death's-head Hawk-moth in Scotland was only a photograph. Arriving in the primary school playground to collect my children, I was immediately intercepted by another mother who, knowing my reputation, was keen to show me a picture of the 'enormous' moth she had photographed earlier on a local tennis court. She and her tennis mates had no idea what it was, but were too scared by its size to move it so had been forced to abandon play. If only she had known how gladly I would have diverted my day to see this moth in the flesh, and to relocate it from the court for them.

A few years later another one turned up in East Lothian, though unfortunately this one was found dead. It was carefully contained and eventually passed on to me. It now resides in my freezer and I occasionally bring it out to impress people. One winter I lent it to a sixth-form art student for her moth-inspired fashion project. She named it Abraham, a name which has stuck: we are now a household with an Abraham nestling among our frozen supplies of peas and chips. I hope, third time lucky, my next encounter with a Death's-head will be a living one.

Around the world there are other moths which choose to get their sugary fix from fruit. The so-called fruit-piercing moths specialise in this using their mouthparts to probe into soft fruit. Serrated edges adorn the sides of the proboscis and once inserted into the flesh it is shifted up and down, the edges slicing and pulping up the fruit into smoothie mix. This feeding action can cause significant damage to commercial crops; one of the rare cases of the adult moth being a worse pest than its caterpillar.

One of the most widespread and problematic is the Common Fruit-piercing Moth, *Eudocima phalonia*. This moth is currently found through large swathes of Africa, southern Asia and Oceania, but it has invasive tendencies and there is concern that it and its undesirable feeding habits will spread further afield. Their stout barbed tongue has no problem piercing the tough skins of citrus, banana, mango, lychee, pear, grape and more. The damage they cause is made even worse when the vacated puncture hole is infected by fungal or bacterial disease.

Another group of fruit-piercing moths, living in regions of Africa, Asia and Europe, includes species that occasionally turn their barbed mouthparts on animals. These are the vampire moths, *Calyptra sp.*, which pierce their way through animal skin to get a drink of blood. Documented victims include buffalo, elephant and other large mammals. Inquisitive researchers boldly offering a finger to a captured vampire moth report soreness and a red mark left by the piercing, but no lasting damage. Their habits offer them notoriety, but only male vampire moths drink blood, and only occasionally, probably for the salt. For the rest of their lives they, and the entirely frugivorous females, subsist on a less controversial diet of fruit juice.

Needless to say, naturalists have learnt to exploit a moth's sweet tooth and entice them from the night with home-made treats; a technique known as sugaring. Entomologists first reported sugar baits could be used to attract moths in the early decades of the nineteenth century when moths were noticed trapped in bottles filled with sugar and water or beer, primarily designed to control wasps. It is hard to assign the subsequent development of this 'discovery' to a particular person or time; from around 1830 various entomologists were documenting moths being attracted to sugary liquids left out by humans.

In London, brothers Henry and Edward Doubleday were working in their family grocery when they noticed that moths were attracted to the sweet remnants in emptied sugar casks left outside the store room. This gave them an idea. Hefting some empty casks into the nearby country-side, they waited until dusk fell. It was a great success. Many species came to the sugar-crusted barrels. Some, they noted, were 'quite uncommon'. Around the same time, hundreds of kilometres away in Northumberland, P. J. Selby used empty skeps, still containing traces of honey, in a similar way.

Dragging sugar-crusted barrels into the countryside isn't easy, and soon a less cumbersome approach emerged whereby the sugary bait was first mixed up in a small tin and taken to the habitat of choice to be painted onto suitable surfaces such as posts, trunks or vegetation. This is the sugaring technique that is still used today. There are various bait recipes but the key ingredients are usually treacle, dark brown sugar and beer, all heated together into a sticky potion. Depending whom you ask, over-ripe fruit, rum, ethyl acetate ('pear drops') or honey are recommended additions, though there has been little science to demonstrate any superior performance from including these. Purists simply stick with sugar and beer.

One late August I arrange with a friend to try sugaring at a local nature reserve. The evening is warm with barely a breath of wind; unusual conditions for the coast and perfect for

Angle Shades
feeding on a sugar-painted post

moths. Meeting an hour or so before dusk, we make our way across the reserve, stopping to adorn fence posts with the sticky mixture we have each prepared earlier. He is on the beer and treacle. To provide an alternative menu choice I have gone for a claret-coloured combination of sugar and the cheapest red wine. The moths don't seem too discerning and half a bottle mixed with a large bag of white sugar works as a drippy paint or better still as a potion to impregnate sections of rope or cotton sheet which can be suspended from branches. Sometimes, while I am out in the night attracting moths with my sweet decoction, the half-used bottle gets finished by my husband at home. Both of us contented; money well spent.

Posts painted, we take a pause on the edge of the adjoining golf course, lying back on the tightly mown fairway to stare up at the darkening sky and just-appearing stars. It isn't completely quiet – there is the gentle lapping of the tide on the beach a short distance away and a distant, less musical, hum of traffic which I try hard to block out – but it is peaceful. I'm little more than eight kilometres from my bustling household with its slamming bedroom doors and stomping feet, yet I feel like I could be 800 kilometres away. The air is warm, the ground only a little bit damp. I'm sure I could fall asleep here.

But we have sugared posts to check and I have promised I will be home for the night. So, easing out of my comfortable reverie we wander back the way we came, shining dim torches on our sugared posts hoping to find some customers. We are not disappointed. Almost every post hosts at least one busy drinker, dabbing at the sticky sweetness: Autumnal Rustic (*Eugnorisma glareosa*), Yellow-line Quaker (*Agrochola macilenta*), Green-brindled Crescent (*Allophyes oxyacanthae*), Angle Shades (*Phlogophora meticulosa*). Most numerous are Square-spot Rustic (*Xestia xanthographa*), a rather plain brown moth to describe but impressive to watch arranged in random formation on a post, busily guzzling at our offerings. I had long dismissed Square-spot Rustic as ordinary and ubiquitous, but this evening I learn they are more worthy of attention. Seeing them in numbers allows me to appreciate the variation in colour and pattern of their brownness. Soon

I get skilled at recognising their shape and features, easily separating them from a similar brown moth, the Small Square-spot (*Diarsia rubi*), also there but in lesser numbers. In the end we count over five hundred moths and eleven different species; perhaps not as big a variety as a light trap would have caught but almost certainly higher numbers.

Sugaring is a notoriously fickle technique but there is a joy to being out in the night, scrutinising sweetened posts with a torch. On evenings when the moths must be busy elsewhere, there are other creatures to admire. Slugs, wood-lice, earwigs and beetles slide or crawl out of the undergrowth to take advantage of the unexpected free-for-all. I've watched spiders take on flies; seen the face-to-face mating of harvestmen; been obliged to step aside as a badger trundled with blind purpose down my path. Sugaring is also less intrusive than a light trap. It doesn't detain the moths for long. They are simply fuelling up before disappearing into the darkness to disperse, find mates, lay eggs. All while trying to avoid the attentions of the hungry predators of the night.

CHAPTER 9

Evasive Action

Now air is hush'd save where the weak-eyed bat,
With short shrill shriek flits by on leathern wing

'Ode to Evening', William Collins

As children, at dusk, my siblings and I would throw tiny stones into the air in the field behind our house, giggling with mischievous delight as hunting bats swooped in on false promises. More recently and more usefully I have helped survey these tiny mammals. I've climbed ladders for routine checks of bat boxes high in trees. I've joined surveys deep underground, to count hibernating bats in their mid-winter slumbers. A few years ago, I helped my friend Dave monitor a summer roost in a ramshackle building on the outskirts of Edinburgh. The building was poised for development, but first we needed to document its usage by bats.

We arrive a little before dusk to take our positions around the building. I am handed a clicker counter, shown the likely exit point in the eaves above and left on one side of the building with instructions to click in every bat that flies over my head. Sounds easy enough. Dave takes up his position on the other side of the building to do the same. Already I can hear a faint chirruping is coming from somewhere within the roof space, adding to my anticipation of what is to come.

Alone, I wait in the deepening dusk, surrounded by brambles and rubble with the noise of traffic coming and going unseen in the distance. I'm not sure quite what to expect but keep my eyes fixed on the roof and my thumb poised in readiness over the counter, occasionally trying a click to check I have the correct technique, before zeroing it again. The squeaking noises are getting louder and then I think I make out a small shape flapping fast as it leaves the eaves. Was that a bat? Another one follows, then three; definitely bats heading out with purpose. Click, click, click, click, click. Very soon they are streaming out almost faster than I can click. One after another they come, like bubbles being blown from a hooped wand, impossible to see in any detail but just amazing that all these small mammals had been roosting together within the building's roof space. Their flapping is just audible before they disappear into the night to do whatever bats do; swooping on flies, small moths and anything else edible in their path. Every summer evening at roosts all over the country bats will be heading out to do the same, and every dawn will see them swarming back home, yet I had never before experienced this short episode of wildlife wonder. In about ten minutes it is over. I fear my count ends up rather more approximate than exact.

Biologically, the saying 'as blind as a bat' holds little sway; nocturnal bats have eyes and can see, but they don't rely on vision to hunt. Instead, these bats hunt with sound, using echolocation. As they fly, they emit a pulsing stream of ultrasonic squeaks which emanate into the dark as sound waves, bouncing off the objects they hit. The returning echo is detected by the bat, allowing it to discern obstacles and snacks in its path. Details of size, speed and direction are picked up, allowing them to expertly home in on suitable prey.[1]

Every time a night-flying moth takes to the air, it risks detection by an echolocating bat. Even moths minding their own business on leaves or other surfaces aren't necessarily safe. Some bats are adept at noticing unusually echoing leaves or picking up on tiny rustles as a moth shifts position, swooping in to pick the tasty morsel off, a feeding behaviour known as gleaning.[2] Over the millions of years that bats have been hunting moths, the two creatures have been trying to outwit each other; the bats developing ever more sophisticated ways to catch moths and the moths countering this by refining their avoidance strategies.

If a moth can hear a bat coming, it can dodge out of the way by changing course or effecting a sudden and unpredictable free-fall towards the ground. About three-quarters of moths are able to hear. In most species, the ears are specialist structures called tympanic organs, positioned just behind the head. These are essentially a small opening with a thin membrane inside, which acts like an ear drum. Sound vibrations are picked up by nerves positioned behind the membrane, which pass the message on, to be interpreted by the brain. Tympanic organs are the original moth ear, the design that is found in the most primitive eared moths, but other designs have since evolved. Hawk-moths for example listen with their mouths, having repurposed part of it into a thin-walled resonating chamber.

Whatever the design, the 'ears' are particularly good at detecting the high-frequency sounds of hunting bats. Because of this, scientists once assumed that they must have evolved specially to listen out for these night-time predators. Recent research has revealed this is not the case.

Akito Kawahara and his team at the University of Florida undertook a huge and complex study comparing the DNA of many different moths.[3] They wanted to understand when in the evolutionary history of Lepidoptera different characteristics first appeared. By looking at the genetic make-up and features of currently living moths alongside accurately aged fossil moths they discovered that moths with ears were flying around their prehistoric worlds at least 28 million years before echolocating bats were on the scene. These ears must have once been used for something other than bat avoidance, probably to hear other approaching predators but perhaps also for communicating with each other. Once bats started hunting nocturnal insects using echo-location, it is likely moths adapted, or retuned their ears to better detect the ultrasonic frequencies being used to hunt them down.

Some bats echolocate with frequencies that are outside of

the moths' optimal hearing range, giving them a better chance of swooping in undetected. Meanwhile, moths have become skilled at confusing bats, which they do in a number of ways. Depending on the species of moth, this involves a combination of voice, soundproofing and tails.

Some moths make a noise. This might be to sing attractively to other moths for courting purposes but more often it is used to shout at a hunting bat. One group of noisy moths, the tiger moths, have been particularly well studied. They have a sound-producing organ called a tymbal. This is a thin membrane, textured with minute ridges, that emits a high-frequency sound when buckled by flexing the underlying muscle. Originally it was thought that the noise simply jammed the bat's sonar, scrambling the echo the bat hears and confusing where the moth target is, but recent experiments suggest that shouting also works as a recognisable warning, a bit like bright colours, to tell the bat that the moth is unpleasant to eat and best avoided.

Just as a bird learns to recognise a foul-tasting tiger moth based on its bright contrasting colours, bats soon learn to avoid them by recognising the noise they make. Video recordings of foraging bats in an experimental arena have shown that once a bat has had an unpleasant taste experience with a tiger moth, spitting it out after capture, it makes no attempt to swoop in on the same species again. The decision to avoid is made following a noise emitted by the moth. Amazingly, there are other moths that are perfectly palatable, but are able to mimic the sounds made by noxious tiger moths, successfully fooling hungry bats.

Some moths are deaf. They can't hear a bat coming so opt for stealth tactics to slip under the bats' radar. These moths rely on the architecture of their scales to help them evade detection. Fluffy bodies and the fine sculpting of individual wing scales absorb the bat's ultrasonic sounds, stopping them from echoing back to the bat. Fascinating research at Bristol University has shown that if you shave the fluff off a silk moth's body, bats are far better at catching it. Using powerful scanning electron microscopes, they also scrutinised the structure of the wing scales of the moths. The images are incredible. Each scale is nothing like the smooth surfaced tile I had imagined, but instead exquisitely scalloped, textured and ridged. When a wave of sound hits the wing, the scales start vibrating, different sizes and shapes responding to different frequencies in such a way as to make an extraordinarily efficient absorber of sound. That such a thin layer of wing can provide such good muffling qualities means the design could even pave the way for development of better coatings for military submarines or superior sound-proofing in buildings.[4] If it can be replicated in paint or wallpaper form, I'm buying it to cover the walls of my son's bedroom.

Another effective way to avoid becoming breakfast for a bat is to trail ribbony streamers behind you. The Luna Moth (*Actias luna*) of North America has large pale-green wings which almost seem to glow if they are caught in torchlight after dark. With a wingspan of around ten centimetres it is one of the larger moths in North America but even more striking are the long, slightly crumpled-looking extensions to the hindwings. At rest they lend a certain elegance to the moth and in flight they flutter and spin behind it like kite tails. However, these streamers aren't simply there for decoration: they seem to be effective at confusing hunting bats, making it difficult for them to work out where to sink their teeth. American entomologist Juliette Rubin has investigated interactions between bats and different species of silk moth, some with elaborate tails, some without. She shortened or cut off the hindwing tails of some, and glued longer or differently shaped tails onto others. The moths were released into a large arena tethered to a long string which allowed them free flight within a fixed sphere. A hungry bat was introduced and the subsequent combats recorded with cameras and microphones. The convincing outcome was that long tail streamers fooled the bats and saved the moths.[5]

As with so much of moth behaviour and ecology, their battles with bats are waged under the cover of darkness. This makes it challenging for us to see what is going on. Only a tiny proportion of interactions between moths and bats have been investigated and we know very little about how these relationships play out in truly natural settings. How many more of the weird and wonderful wing shapes and body parts of moths might turn out to be cunning adaptations to help avoid detection by bats?

Despite the moths' best efforts, bats still manage to consume them in large numbers from the night skies. But only adult moths are at risk. Younger life stages, the caterpillars, have different enemies. Birds, particularly those with a nest of chicks to feed, can get through large quantities, and reptiles, small mammals and predatory invertebrates all enjoy caterpillar snacks too. But perhaps the most fearsome of all is a group of adversaries that are always smaller than their prey. They don't simply gobble their victim whole, but let their young feed off the still-living body, killing it gradually over a period of time. These are a diverse, endlessly fascinating selection of insects with ingenious lifestyles, known as parasitoids.

Parasitoids lay their eggs in, on or near another organism, which becomes known as the host. The hatched larvae feed and develop either in or on this host, eventually killing it. (This is slightly different from a parasite, which typically lives and thrives on its host without directly causing death.) Several insect groups have opted for the parasitoid mode of life, including a few species of moth,[6] but it is particularly prevalent among the wasps and some flies. A wide range of invertebrates can become unwilling hosts. In moths, the eggs and pupae are sometimes attacked but most often it is the caterpillar stage that is targetted.

My first knowing encounter with parasitoid wasps was with dead ones when, many years ago, I volunteered at the National Museum of Scotland, helping wasp expert Mark Shaw curate some of the minuscule specimens in his care. Lined up in columns, their tiny black forms, stuck to small triangles of card, belied the damage they had once caused their far larger hosts and gave no hint of their fascinating lifestyles.

My task was essentially one of writing labels in the tiniest, neatest script I could manage and attaching them to the appropriate specimen. In the process I learned a little of how they lived their lives, but it is only much more recently, largely through my increasing familiarity with moths, that I have discovered more of the range of ways these wasps ply their parasitic trade. They are a group of insects which I readily admire, even though my beloved moths rarely come out well from the interaction.

Nearly twenty years after the encounter with museum specimens, I was conducting plant surveys on some coastal grassland. I get easily distracted when surveying plants. My required attention to the hairiness of the stem, the shape of a ligule or the colour of a sepal is frequently lost to the buzz of a bee or the scurrying of a spider among the leafy thatch. Today I notice a white fluffy growth on a grass stalk. It doesn't look like a fungus or some other plant disease and I wonder what it might be. Adopting the classic pose of botanists and entomologists everywhere (elbows on the ground and bottom in the air), I peer more closely. Now I can see that it is actually a collection of small white capsules, each no more than a millimetre or two long, clustered loosely but firmly around a two-centimetre section of stem. They look like tiny silken cocoons. Then I notice the stem isn't quite what it first seemed. There is a green caterpillar under those white clusters. It looks fresh, but I can't decide if it is actually alive. I'm not sure of its exact identity either; in a different life it would be on its way to becoming one of the White butterflies (Pieridae). I carefully pick the grass

stem and put it into a small pot. I want to find out what is
going to happen next.

Later, an internet search tells me that the small white
capsules are the cocoons of a parasitoid wasp from the
Braconidae family. There are likely suggestions of exactly
which species, but I don't know if I have enough information
to make a definitive identification. I put the pot and its pupae
to one side.

Less than a week later and my pot is suddenly alive with
tiny, fidgety, black wasps. At a glance they might easily be
mistaken for minuscule flies, but the body shape, the
waving antennae and the presence of four, rather than two,
tiny wings prove otherwise. Under a microscope they are
actually rather smart-looking insects.

A few weeks previously, the mother wasp would have laid
a batch of eggs inside the body of the happily munching
caterpillar. Once hatched, the developing larvae feasted until
they were fully grown, saving the vital organs of the caterpillar until last. Satiated, they worked their way to the
outside of the caterpillar body, each finally spinning a silken
cocoon in which to pupate and complete its development
to adulthood. The caterpillar breathed its last. Not a lifestyle
for the squeamish.

How do these parasitoid wasps locate suitable hosts, and
what can the caterpillars do to protect themselves from
attack? In exactly the same way that a male moth detects a
calling female moth, the wasp's constantly twitching
antennae are sampling the airwaves for giveaway scents

Listrognathus obnoxious ovipositing
into a Six-spot Burnet cocoon

which might guide it to potential prey. As any shy person knows, the best way to avoid encounters with others is to hide from view and keep quiet. However, these strategies offer little protection from predators that seek out their victims by smell. If it is to avoid attack from an egg-laden parasitoid, a caterpillar needs alternative strategies. Being odourless would be a good start.

Even if caterpillars could mask their own scent effectively, the actions of living still leave an unavoidable trail of volatile chemicals in their wake that parasitoids have learnt to take advantage of. If you pluck a leaf and rub it between fingers and thumb it smells. The scent is a combination of different chemicals that the plant releases when its cells are damaged. In a similar way, as a caterpillar munches a leaf, the damage causes signature chemicals to waft into the air. These are in concentrations undetectable by our relatively insensitive human noses, but to a parasitoid's antennae the signal is loud and clear. When a female gets a whiff of damaged plant she homes in to investigate further. If the damage is being caused by one of her favoured host caterpillars, she has found a place to lay her eggs.

The hairs and spines on some caterpillars offer a degree of protection by making it harder for a parasitoid to lay its egg in or on the body, and caterpillars that live within trees or in the soil have a physical barrier which prevents direct attack, at least from most airborne aggressors. Not to be outwitted, some wasps have impressive egg-laying equipment; the abdomen tip is like a long hypodermic needle which they can insert through spiny hairs or solid surfaces to get their eggs into the soft flesh of unsuspecting victims beneath.

Caterpillars living above ground can also try evasive action when under attack, wriggling and writhing to make it hard for the parasitoid to get a grip, or they can attempt a getaway with a sudden drop to the ground. Caterpillars also have an immune system to help protect them.

The immune response is capable of engulfing and killing foreign invaders and this includes parasitoid eggs or very small larvae in their body. But it isn't foolproof; some parasitoids go a step further. They can coat their eggs with substances that render them undetectable by the caterpillar's body until it is too late. Even more cunningly, some parasitoid wasps have developed ways to overcome the caterpillar's defence by teaming up with a virus. When the female wasp injects her eggs into the caterpillar host she also injects virus particles. The virus effectively disables the caterpillar's immune system and the wasp eggs hatch out and develop without assault. In return for its help, the virus gets a chance to do some of its own replicating.

Parasitoids might appear to be an invincible enemy of moths, but they are a crucial part of ecosystems and have an important role in regulating the size of moth populations without eliminating them. Too many caterpillars could consume all their foodplant, leading to mass starvation and local extinction. Parasitoid attack helps reduce the number of moths reaching adulthood. This leads to fewer caterpillars the following year which in turn means fewer hosts for the parasitoids to find. As a result, the parasitoid numbers drop, allowing moth numbers to build up again. The result is a regular fluctuation in moth numbers over the course of years, tracked by a fluctuation in abundance of its parasitoids. This is particularly apparent when the parasitoid is dependent on just one host moth species, as many are.

Just as with adult moths and bats, there is an ongoing and complex evolutionary battle between caterpillars and their parasitoids. Each one is forever trying to get one-up on the other, yet neither ever claims absolute victory.

In the end though, it is perhaps the moths that have the last laugh. Parasitoid insects need a host to survive. They will always be rarer than the creatures they prey upon. When considering conservation of scarce moths, it is worth remembering that there might well be an even more unusual, but just as important, parasitoid fly or wasp, whose survival is irrevocably linked to that of the moth.

Due to their relative rarity, covert lifestyles and the fact that few entomologists specialise in them, parasitoid insects are generally poorly known. There is still much to discover about their behaviour, the hosts they use and the impact they have. Moth enthusiasts may not be fond of the way parasitoids live their lives, but they can contribute to understanding them better.

Rearing a caterpillar, that is keeping one you find until it pupates and becomes an adult moth gives a unique and privileged insight into the life of different species. From the Pimpinel Pugs living on my nearby cliffs, to Heralds hiding in my local culvert, looking after caterpillars as they develop has provided a fascinating way for me to get to know moths better. As with any venture involving responsibility for another life, it requires a bit of commitment; they need to be supplied with suitable food and their cage must be kept clean. On the other hand, no special equipment is needed and it is perfectly acceptable to return the captive back into the wild if you find you can no longer look after it. Once admired, the newly emerged adult can be released back into the place its caterpillar was found, to complete its life.

I have a selection of repurposed margarine tubs, lidded with scraps of net curtain and rubber bands, which are used to accommodate my temporary caterpillar guests. Whereas once this was a family activity to enhance my children's fascination in the natural world, now my caterpillar rearing is mainly selfish; for my own enjoyment and education. I learn how they change in looks after each moult, what leaves they particularly like, how long their growing takes and, with luck, the winged moth they eventually become. Newly emerged, pristine wings just dried and unblemished, a virgin moth is extraordinarily beautiful; a mint condition that is rarely encountered in the wild.

Sometimes, however, this beautiful insect never appears. Instead, a fly or wasp is suddenly there, urgently wandering about the cage, a hole in the exited pupa the evidence of what has happened. Tempting though it is to throw the lot in the bin in despair, this would be a waste. The unwanted parasitoid can be retained. Its identity will probably provide more information than the expected moth would have done; it might even show an association that was not previously known or turn out to be a new species of insect for the region.

Roy Leverton tells a story of a brood of fly maggots that he once stamped on in frustration, when they emerged from a pupa he was hoping would become a large and magnificent Pine Hawk-moth (*Sphinx pinastri*). Fortunately, he didn't squash them all, and having calmed down a bit, he sent the remaining unwanted invaders to an expert for identification. They turned out to be *Phryxe erythrostoma*, a tachinid fly that relies on hawk-moth caterpillars and was, excitingly, a new species for Britain.

I send my uninvited wasps to Mark Shaw. With a neat label detailing what it is and where it came from, he adds the specimen and its exited pupa to the collection he compiles for the National Museums of Scotland. It's the very same ever-expanding collection of wasps that I happened to work on all those years ago.

CHAPTER 10

Masquerade

I shall not fail, though, to detect
The four lovely gauze wings
Of the softest Geometrid moth in the world
Spread flat on a mottled pale birchtrunk.

The Gift, Vladimir Nabokov

I am in the middle of a bog, almost in the middle of Scotland. A kilometre or two away as the golden eagle flies there is a plaque on a large boulder, announcing the exact geographical centre of the country; apparently more or less the point at which a cardboard cut-out of Scotland could be perfectly balanced on the tip of a pencil. From my soggy hillside I look northwards, down on the silvery ribbon of the River Spey as it snakes through the flood plain below. Beyond, the horizon is decorated with mountains, today rendered in hazy shades of green, purple and grey. Around me are heathery hummocks, interspersed with small stands of trees, scattered boulders and damp mossy cushions. Wetter, craggier and wilder than the moors near home and with plenty that is unfamiliar to divert me.

Sunning itself on a broad leaf is a striking hoverfly, *Sericomyia silentis*, striped in black and yellow and looking every bit like a wasp. It doesn't seem bothered by my presence, almost as if it is enjoying my admiring scrutiny. I come across an abandoned shred of reptilian skin, hidden in a heather tussock and presumably shed from a common lizard. I finger the papery fragments, wondering how far away its former occupant is lurking; probably not very. Buzzing between the purple pompoms of devil's-bit scabious flowers, a delightfully furry tan-coloured bumblebee is fuelling up on nectar. Common carder I think, but it buzzes off before I have time to interview it properly for a more certain identification.

Moths are proving frustratingly elusive. I uncover a hairy black caterpillar among some heather, with tufted white spots and a line of scarlet running down each side of its body. It is unfamiliar, so I check its identity in my guide-book. The book says Light Knot Grass (*Acronicta menyanthidis*), an exotic species for me; its British distribution mostly restricted to boggy heaths and moors. I'd love to see an adult and I briefly contemplate taking this caterpillar home to rear before quickly thinking better of it. Suitable food-plant might be hard for me to replenish locally and what would I do with the new adult? It is a long way to come to return it to the wild.

As I prepare to return to the surveys I'm really here for, I catch a glimpse of pale flutterings in a small gathering of birch trees about thirty metres away. Creeping closer, I see more movement. Certainly moths, perhaps some sort of carpet moth? There are at least half a dozen of them, taking short flights before landing somewhere on the trees. I follow one that appears to land on a trunk about ten metres away. Keeping my eyes on the spot where I thought it had alighted, I creep slowly forward, trying not to trip over tussocks or land my foot in a boggy hollow. Finally, face to face with the tree, I examine the peeling silver-grey bark with its knots, fissures and sproutings of lichen. Mysteriously there is not a moth to be seen. Maybe I had confused where it had landed. But as I reach out to rest my hand on the trunk for balance, a moth takes off under my nose, there all the time and hiding in plain sight.

I see another land and try again, once more fixing the spot in my sights and approaching as stealthily as I can. It is hard to make out, but this time I can see it, very well camouflaged as it rests with wings spread flat against the bark. It's a lovely Dark Marbled Carpet (*Chloroclysta truncata*).

The carpet appellation has nothing to do with what they might eat (most carpet moth caterpillars eat leaves). It arose from early naturalists' fad of naming moths after things around them. Brocades, wainscots, footmen, tussocks and lackeys; their selection has left an eclectic mix. Carpet might seem an uninspired choice of name, but back then carpets were richly patterned and an expensive luxury, so perhaps it is more of a compliment to the moths than it now seems.

Dark Marbled Carpet

Dark Marbled Carpets are variable in colour, with any amount of white, grey, chestnut or chocolate browns marbling in erratic zig-zags across their wings. This one is predominantly white; I wonder, does it 'know' that it has pale wings so it will be best hidden if it rests on the pale trunks of silver birch?

I follow another that hunkers down somewhere in a patch of low-growing bog myrtle. It takes a while to spot, but I persist as I'm sure it is there. This one is even lovelier than the last. The background colour is again quite pale, but this time striking streaks of chestnut brown mixed with bands of pure black undulate across the wings. The pattern reminds me of a marble painting activity I tried with my children when they were small, in which coloured paint is squirted into a tray of white paint and swirled around (very messy, very fun). Or maybe it is like the dribbling pattern around the edge of an over-iced chocolate birthday cake. I spend the next half hour absorbed in perfecting my Dark Marbled Carpet hunting technique and trying to take photographs. But none of my pictures turn out to do the moths' finery justice.

Moth protagonists go to some lengths to showcase the more colourful moths we have, in an effort to emphasise that not all moths are dull or drab. But truth be told, the majority do come in various shades of brown and grey. This doesn't make them boring or lacklustre; their intricate patterns and fine-sculpted shapes are beautiful, matching perfectly the wood, stones and vegetation where they rest in stillness during the day. Mossy greens, conker browns, cream, chocolate and deepest black; add stripes, swirls and splotches and you have some of the most striking wildlife Britain has to offer. Scalloped Shell (*Rheumaptera undulata*), Mocha (*Cyclophora annularia*), Streamer (*Anticlea derivata*), Waved Umber (*Menophra abruptaria*), even their names are enticing. Without help from a light trap though, finding them is difficult, for their beautiful patterns provide excellent camouflage. These moths are trying to hide.

Daytime predators, notably birds, tend to hunt by sight. Moths that rest openly, for example on tree trunks, have lines, speckles and blotches in darker or lighter shades, suggestive of bark fissures, lichen patches and other textures. They blend in perfectly with their surroundings and evade detection. Wavy wing edges and bold blocks of colour help break up any tell-tale mothy outline and further the trickery.

As natural substrates such as tree bark or rocks are variably patterned, so too are the colours and intensity of wing markings of moths that want to hide on them. This allows each moth to choose the best place to hide and also makes it harder for birds to learn a consistent search image to look for. Even moths of the same species can look very different, and moths living in woodland with predominantly pale-barked trees will tend to have lighter wings than the same species living in a wood with darker trees. For the moth-er, this variation causes frustrating identification challenges – how can two moths with such different colourings be the same species?

Camouflage is an important survival strategy. Moths that don't blend in are more likely to be spotted by predators and eaten. These misfits don't survive to breed and the genes that code for that colour or pattern are removed from the population. This is the process of evolution by natural selection, first published by Charles Darwin and famously demonstrated in now legendary experiments on the Peppered Moth (Biston betularia) in the 1950s.

The Peppered Moth is an impressive moth with stocky body and hawkish wings spanning about five centimetres. Most individuals are a whitish colour covered with a multitude of black spots and flecking, making them look like they have been sprinkled with black powder. By day, the moths hide from sight by blending in, their wings particularly well camouflaged against lichen-encrusted trunks and branches.

Nineteenth-century Britain was in parts a grimy place. During the Industrial Revolution coal-powered factories pumped plumes of sooty smoke into the air. Where the soot settled, surfaces turned black. Trees, walls, buildings, especially those close to or downwind of the factories, were coated. With the soot came noxious gases that killed the lichens which had for centuries been doing their slow growing on these surfaces. The pale Peppered Moths were no longer very well concealed by day and became easily found pickings for hungry birds. Fortunately, the Peppered Moth has a dark version, a form known as 'carbonaria', with entirely black wings. In fact, the wings look like they have been thickly coated with soot rather than lightly sprinkled with pepper.

The first British record of one of these dark forms came from Manchester in the middle of the nineteenth century. About fifty years later nearly all the Peppered Moths being found in the area were the dark form; the whiter version was rarely seen. In 1896, the Victorian lepidopterist, J. W. Tutt, put forward an explanation of the rapid rise of carbonaria, suggesting the dark-coloured moths found themselves better disguised from predators in polluted areas and thus survived better than the paler forms.

Tutt's hypothesis wasn't widely accepted. Many scientists didn't believe that being camouflaged by day was sufficient protection against predators for a night-active moth. Others thought temperature or humidity, or caterpillars ingesting pollutant residues on leaves, might be responsible for the darker moths becoming the dominant form. It wasn't until the 1950s that the theory was finally put to the test by the brilliant natural historian Henry Bernard Davis Kettlewell.

Kettlewell trained as a medical doctor, but his real interest and passion was for natural history and in particular moths. With little enthusiasm for general practice, he must have been delighted when, in 1952, he was taken on by academics at the University of Oxford to carry out fieldwork on the Peppered Moth. The objective of the research was to provide real-life evidence for Charles Darwin's theory of natural selection and to see if, as Tutt had suggested, dark moths were better protected from bird predation in polluted areas.

Kettlewell was a devoted scientist and went to extraordinary efforts to rear huge numbers of both colour forms of the moths and observe them in the field. His experiments showed that birds did eat Peppered Moths during the day and that they preferentially consumed the moths that didn't match their background. This meant the dark forms were at an advantage in sooty industrial areas and became the more common form. In unpolluted rural areas the paler more mottled form dominated.

So neat were Kettlewell's experiments that they quickly became text-book examples of natural selection and evolution in action. I can remember learning about them at school. There are examples from other animals, but none as easy to visualise and understand. Studies on moths carried out elsewhere in Britain, the Netherlands and America gave similar results.

Even so, Kettlewell's results were later called into question.

Surely they were too perfect to be true? Kettlewell, by now no longer alive to defend himself, was accused of inadequately designed experiments and fudging his results. Large egos got involved, personalities came under unjustified and unpleasant fire. Creationists jumped in, cherry-picking facts and publishing them to suit the stories they wanted to tell. As details became more distorted and publications quoted without question or context, it became hard to work out what was fact, error or fabrication.

The matter was finally cleared up once and for all in 2008 by Cambridge academic Michael Majerus who systematically and carefully ran a number of experiments designed to address all criticism. He showed, just as Kettlewell's experiments had shown, that the differential selection by birds, based on their ability to notice the moths, sufficiently explains the prevalence of different forms of Peppered Moth in different areas.[1]

Kettlewell's legacy also lives on in his moth collection. Amalgamated with those of the great entomologist Edward Alfred Cockayne and the enormous collection of Sir Walter Rothschild, they now form the core of the British and Irish Lepidoptera collection at the Natural History Museum in London. Excitingly, as part of the ongoing efforts to digitise museum collections, the Rothschild-Cockayne-Kettlewell (RCK) collection can be viewed by enthusiasts and researchers the world over through an online data portal.[2] Among the many thousands of specimens that have been photographed alongside their labels are over 2,500 Peppered Moths. Here, with a few clicks of your mouse, you can see some of the variously coloured individuals caught or bred by Kettlewell himself.

Blending in by trying to look the same as your background is just one approach to foil would-be predators. Another is to masquerade as something else. No need to hide, just pretend to be something that isn't worth eating.

Many moths mimic wood or leaves. Take the Buff-tip (*Phalera bucephala*), a fabulous stick lookalike, two pupae of which my grandma once gifted me. At rest it wraps its greyish wings around its body, giving it a cylindrical shape, the dimensions of a small stubby stick, about three centimetres long. Different shades of grey and fine black accents render the wings quite bark-like but the pretence is considerably enhanced by a pale brown head and matching brown circle at the wing tips. These patches look just like freshly snapped exposed wood, perfecting a broken-off bit of twig masquerade and allowing the moth to sit, undetected, in plain sight.

Red Sword-grass (*Xylena vetusta*) is another moth that rests with its wings folded close to its body. Its streaky crinkled pattern gives it an uncanny resemblance to a small piece of rotting wood. To further the ruse, in the bottom of a light trap they tuck their legs in and 'play dead', rolling around as the trap is tipped, behaving just like a fragment of wood. I have a friend who is a botanist and he likes to rile me gently by suggesting plants are far superior to and much more fascinating than moths. Yet he conceded there were exceptions when I showed him photos of a selection of Red Sword-grass I had caught.

Imitation need not be restricted to looking like vegetation. A surprisingly frequent ploy, adopted by various small animals, is to pretend to be a bird dropping, for no self-respecting predator would deign to eat excrement.

As you might expect, bird dropping mimics have white, grey and black as their colour scheme, often with fawn brown highlights, arranged in patterns that are remarkably realistic. In the UK we have a lovely little moth, wings not much longer than a centimetre, called Chinese Character (*Cilix glaucata*). Its unusually smooth curved profile and apparently glistening olive-grey central blotch make its disguise impressive.

As with all long-coveted species of moth, I can readily recall the moment I eventually got lucky and had my first and, as yet, only encounter with a Chinese Character. I had joined a gathering of entomologists for a weekend in southern Scotland. From Friday evening to Sunday afternoon I was at liberty to think of little more than insects. It was an opportunity to meet fellow moth-ers from across Scotland, to put faces to email addresses, and of course to go out by day and night to search for new and exciting moths. I didn't get much sleep but so total was the break from my normal life that back home in the days afterwards it felt like I had been on holiday for at least a week.

The Chinese Character was almost the last moth I saw that weekend. I had reached my final light trap on the final morning, and there sitting on top of my trap's lid was a neat little moth, legs and antennae tucked out of sight, smaller than I had imagined but unmistakeable. Quite unlike any moth I had met before, and even more like a small bird dropping than I had expected. I later learned that the lids of light traps are a common place to find Chinese Character. So effective is their feculent disguise that they are often the only moth remaining around the trap after breakfasting early birds have departed.

The unappetising nature of bird droppings is a global phenomenon. As well as adult moths, there are caterpillars and a selection of other animals masquerading as avian excrement worldwide. In North America the rather lovely-named Beautiful Wood-nymph (*Eudryas grata*) sometimes also going by the rather unlovely appellation 'bird-poop moth', is from a different moth family yet remarkably similar in looks to our Chinese Character; just larger and with a somewhat more rugged profile. I was amused to read a social media post describing how somebody had rushed into their garden with a camera on sighting their first ever wood-nymph moth. They were even more excited when it stayed beautifully still, allowing for some lovely photos to be snapped. After the photoshoot it remained in perfect bird-dropping disguise until a gentle poke revealed that it wasn't actually a moth at all; just a now very well-photographed splodge of bird excrement.

However, the bar on dung mimicry reaches its highest level with a species from Southeast Asia closely related to our Chinese Character; the incredibly decorated *Macrocilix maia*. This moth has delicate white wings which are held open at rest like a butterfly. Spattered across the middle of both hindwings is a pale grey and brown blotch with convincing glistening highlights created by streaks and spots of white. A liquid bird splat for sure. But then, next to this splat on each forewing are matching marks that take on the appearance of a pair of red-eyed flies, positioned as if feasting on the bird dropping. The moth even emits a pungent aroma to further its disguise and convince would-be predators to leave well alone. It has been dubbed the Mural Moth because of this apparent 'picture' drawn across its wings. The image is extraordinarily realistic; however, maybe we humans are getting carried away in interpreting the meaning of the patterns.

Bird droppings may be off-putting, but they are hardly scary. Some moths try shock tactics to deter predators that dare to get too close, suddenly revealing hidden brightly coloured wings. One group of moths, belonging to the genus *Catocala*, are particularly striking, their name roughly translating from Ancient Greek as 'beautiful behind'. We have seven species in Britain, but there are around forty species in Europe and a hundred or so in North America.

The Red Underwing (*Catocala nupta*) is a large moth, which by day rests motionless on tree trunks, its mottled brown-grey forewings blending seamlessly with the bark background. This camouflage is usually sufficient protection from harm, but if discovered they have a hidden defence up their sleeve, or rather under their wings. The drab grey forewings are lifted forward to suddenly reveal their undergarments – a pair of shocking-red hindwings. After the startle, they take off in an erratic zig-zag flight, red wings flashing in a hard-to-track trajectory before settling on another trunk. Once again, their forewings shroud their hindwings and they melt back into their surroundings. Close observation has shown there is more than just colour to their startle. When flying to escape harm, they flap more slowly than during their normal commuting flight. This makes the flashing more apparent and more confusing to a predator in pursuit. At the higher wingbeats of normal flight, the colour is reduced to a more constant, and less noticeable, blur.

Rather than sticking with simple patches of bright colour, some moths raise their scariness stakes higher by incorporating intimidating eye-like markings on their wings.

The loveliest eyespots in the moth world have to be those adorning the wings of silk moths in the family Saturniidae. The family name apparently derives from these ocular patterns: the large spots on the wings of some species are made up of concentric circles, supposedly like the rings around the planet Saturn, though it takes more than a little imagination to see a cosmic connection. The moths are always striking: large-winged with furry bodies and a soft plush look to them, topped with feathery antennae. If a moth were to be made into a soft toy, these would be the ones to go for.

We only have one silk moth in Britain, the impressive Emperor Moth (*Saturnia pavonia*). Both males and females have four striking eyespots, one on each wing, each a dark centre containing a small arc of white highlights, and surrounded by a pale rim. The smaller day-active males usually appear little more than a blur of orange as they whizz with purpose over grass and heather in search of a mate. It's a challenge to keep up with him, let alone notice any scary eyes. Empresses are larger and very different in colouring, with elegant soft grey and creamy whites replacing the male's fiery orange and browns. They are much more sedate and hard to find among heather tussocks, but occasionally turn up in light traps.

Emperor moths are not particularly unusual in Britain; the vivid green caterpillars with rows of pink and black pimples are often encountered on heather moors and bogs in late summer. However, the speed of the males and the secrecy of the females make the eye-spotted adults much harder to enjoy.

Fortunately for those wanting a closer look at these majestic moths, there is an effective pheromone lure that mimics the scent of an Empress looking for a partner. Emperors find this smell impossible to ignore.

One beautifully sunny April day I unilaterally decide my children need fresh air. They were just about old enough to be left at home alone, and they knew it, but this was one of those mother-knows-best occasions when my desire to show them something new, or maybe just prise them from their screens, took priority. However reluctant at the outset, they are nearly always better for an out-of-doors outing afterwards. Also, I thought, viewing Emperor Moths is an experience to be shared and something they were bound to enjoy.

On this occasion reluctance manifested itself in an agonisingly long time to find appropriate socks and shoes, followed by the time-worn argument over who was going to sit in the front seat of the car.

'You sat in the front last time!'

'But I always feel sick if I sit in the back.'

Eventually, with the usual victor in the front seat, we are off, heading south into the Lammermuir Hills to a place I know Emperor moths fly.

Parking in a small lay-by, we make our way down over heather and rocks to a clear tumbling stream. It is a beautiful spot. Last time I came here was with a birdwatcher friend and we enjoyed wonderful views and the beautiful melody of a ring ouzel. Today it is just red grouse cackling, the descending piping of a meadow pipit and the sound of water over stones. But Emperor moths and their scary eyespots will be patrolling somewhere, I'm sure.

I hang the pheromone-infused bung in a small net bag over a low fence and allow it to wave provocatively in the gentle breeze. My young company is less than enthusiastic; one has yet to get out of the car, another has resorted to passing the time by throwing stones in the stream, the other is sitting quietly on her own nearby. Undeterred, I instruct them to keep an eye out downwind, as I predict that is where the moths will come from as they follow the alluring whiff upwind. If they come, that is. I lie back in the heather for a few minutes, close my eyes and relax.

Five long minutes later, nothing has happened. But just as I am about to suggest we take a short walk, or maybe move the lure to another spot, I catch a glimpse of orange. Definitely an insect and certainly a largish one. With a much more powerful and purposeful flight than a butterfly, it must be an Emperor Moth. My excited cry brings the children to attention, and forgetting their display of reluctance, or perhaps just in desperate need of some entertainment, they gather near the lure.

Another Emperor arrives and together the two flit urgently around the net of the lure, investigating a promising-smelling mate that is not a mate. Soon there are three moths jostling, barely settling, any striking eyes on their wings lost in a blur of orange and cream. I catch one in a net and put it carefully into the shade of a large heather tussock, where it will hopefully cool down a little and pause for a moment.

After a few minutes we gently coax the net open and our temporary captive is revealed. The overall impression is orange, but in the detail there are different shades of russet, brief streaks of red, and pale bands of white. We can only see the two upper wings, the eyespots on each striking jet-black circles, rimmed with thin circles of orange and more black. The antennae are wonderful too, short but impressively fronded like a miniature fern stuck on each side of the head. As we admire him his wings start to quiver in readiness for take-off. It won't be long before he is up and away, but for this brief moment, as the forewings spread wider, we get a glimpse of plainer orange hindwings, with their own eye-like circles of black. Then he is off, back to the lure to continue where he left off.

Maybe when the Emperor spreads its wings and two eyes become four, a would-be predator gets a confusing impression it's seeing double. Maybe the extra spots get interpreted as double the threat. Whatever the mechanism, anything that confuses a predator and buys time for an Emperor Moth under attack could save its life. There has been much debate about whether these circular spots act as deterrents because they look like fearsome eyes that belong to an animal much bigger, or whether the similarity to eyes is just our own anthropogenic interpretation of the pattern. Experiments with caged birds show that when spots like this are suddenly revealed the birds do show alarm, but they might just be showing fear of something unfamiliar, rather than recognising them as large menacing eyes. It turns out it is hard to design experiments that will demonstrate convincingly the reasons *why* a predator is scared of something.

Back at the lure, there are now six moths fussing around, coming and going, undaunted by the competition. I wonder if a female Emperor Moth has to endure as much attention as my lure. I hope, expect, she can turn off her attractive scent once she is engaged. Maybe she emits other scents, which usefully repel the surplus suitors.

I put an end to the charade and seal the lure back in its container. The moths swiftly disperse to better things. This small patch of moorland suddenly seems very empty. Forgetting to argue, we return to the car. A successful outing, I think. At least for the humans involved.

CHAPTER 11

Long-haul Travel

Not all those who wander are lost.

The Fellowship of the Ring, J. R. R. Tolkien

'It's just beckoning a lick, isn't it?' grins John, as we gently place the moth we have just lifted from the moth trap onto a lichen-encrusted tree trunk.

We are in a wood on a grey September morning. The moth we are admiring is a Merveille du Jour (*Griposia aprilina*). This is a top moth, its pale green wings strikingly accented in black and white, offering excellent camouflage against lichen and leaves. John's licking reference came because, with a little leap of his imagination and breakfast not yet taken, the wing pattern is reminiscent of mint choc-chip ice cream. Each autumn, up and down the country, moth recorders cross their fingers and wish for their own marvel of the day.

Other moths have brightened up this morning for us too. We have enjoyed the shocking yellow fluff of Canary

Shouldered Thorn (*Ennomos alniaria*) with its beady green eyes and superbly plumose antennae; several Feathered Thorn (*Colotois pennaria*), each a slightly different shade of autumn, from lemon yellow to russet brown, but each with the same tiny thorn-shaped mark etched in the corner of their broad triangular wings; and Black Rustic (*Aporophyla nigra*), elegantly clothed in gothic velvet splendour.

All these moths live out their lives, generation after generation, from egg to caterpillar to pupa to adult, in the same area. Merveille du Jour caterpillars eat oak leaves; a long-lived food source that should reliably remain in the same place for centuries. I find this beautiful moth in this oak-rich woodland every year, its colourful presence a reassuring sign that nature's calendar is progressing as it should.

Many species of moth happily complete their life cycle in a single area, as long as conditions stay the same and their habitat remains intact. They may have to range a little to find a mate or seek out suitable foodplants to lay their eggs, but this shouldn't be far. Monitoring these resident moths gives us a handy way to check up on that particular habitat; stable numbers indicate that all is well.

Within populations of these stay-at-home moths, there will always be some itchy-footed individuals eager to disperse further if they can; this is how populations mix and how species spread into new areas. The ability to seek out a new home is particularly important if habitats are destroyed or environmental conditions change, but it can also be advantageous to move to where there is less competition for resources or fewer enemies.[1]

For day-active butterflies and moths, hedge betting is a common strategy for females; the first batches of eggs are laid near where they emerge and mate, and the remainder are saved for a new location if they can find one. Males, on the other hand, are likely to rove as much as they can to catch up with as many females as they can. The habits of nocturnal moths are probably similar, but it is hard to follow them under the cover of darkness, so much of what they get up to in natural settings remains unknown.

In some species though, travelling goes far beyond irregular or opportunistic dispersal. With their sights on far flung horizons, long-distance journeys are part of life for these moths.

We enjoyed the colourful selection of moths gracing my light traps this morning to a wonderful backdrop of 'wink-wink' calls of pink-footed geese, as they moved through the skies above in v-formation. A few days ago these birds might have been nibbling at plants on a mossy lava plain of Iceland, now they make touchdown in Scotland. Each year I look forward to the first arrivals; a reliable indicator of the changing season.

For an individual pink-footed goose, which can live twenty or thirty years, their annual back-and-forth journey between breeding and wintering grounds must become familiar. It is a long way, but by doing it they are maximising their reproductive output. Even if they could withstand the freezing temperatures, there is little for them to eat during the long iced-over Arctic winter. Here in Britain winters might be more pleasant, but the competition for space and resources in the summer months is high. Nests would frequently fail. The geese's solution is to breed up north, winter further south.

Annual migration is a feature of many animals' lives. Birds might be our best-known migrators, but whales, wildebeest and salmon are among the many non-avian creatures travelling long distances back and forth to take advantage of the best conditions at different times of the year. Instinct drives them to make these journeys; something in their genetic makeup means they can't help it. Although I know some of the science behind migration and how it can be explained by a complex mix of magnetic fields, the sun, stars and genes, to me the phenomenon remains an extraordinary feat of navigation and endurance.

Even more extraordinary is that despite their small size and apparently fragile nature, many insects, including moths, make similarly long distance journeys over land and water. Moths have been found on oil rigs in the middle of the North Sea. They have been noticed pausing for a rest on ships crossing oceans. They flap their way through mountain passes. Although weather plays a part, it isn't simply a case of being blown wherever the wind takes them; these moths have some sort of inbuilt compass that enables them to set their initial course and stick to it. Just like their larger avian counterparts, migrating moths journey to seek better conditions and maximise their reproductive potential.

In zoology, migration is usually defined as long-range, undistracted movement in a more or less straight direction to exploit more favourable conditions elsewhere, followed by a return journey sometime later. Pink-footed geese are a good example, with each individual goose making the same annual migratory journey many times in its life. For moths, with short adult lives, the return is almost never accomplished by the same individual but instead by their offspring. In this case it is populations of moths, rather than individuals, that complete a seasonal back-and-forth journey.

This means migrating moths are naïve; they've never done the journey before and will never do it again. With no experienced others to follow they rely entirely on instinct to know when and where to go. Environmental cues of temperature, other weather patterns and day length interact with the moths' genetics to make this work.[2]

In Europe, over the course of each spring, billions of moths take to the skies to leave the hotter dry conditions in the Mediterranean region and travel northward where a more abundant supply of fresh vegetation will feed their caterpillars. They fly high, hundreds of metres above ground, choosing the altitude with most favourable wind speeds. Incredibly, thanks to radar technology, scientists can get a look-in on these journeys and track their routes and progress. Some species can even be identified, based on the size and shape of their radar image and how fast they flap their wings.

Among the more abundant moth migrators to Britain are Silver Y (*Autographa gamma*), Delicate (*Mythimna vitellina*), Scarce Bordered Straw (*Helicoverpa armigera*) and Vestal (*Rhodometra sacraria*). Average speeds are around 50 kilometres per hour, but with a good tail wind speeds of up to 100 kilometres per hour have been clocked. This means a trip from southern Europe only takes a couple of days.

Instinct, refined by thousands of years of evolution, ensures long-haul moth travellers choose their moment of departure well. It would be foolish if not hopeless to take off with a strong headwind, so the moths wait until the wind direction is in their favour. Ideally this will be such that it will carry them to the place they want, but studies on Silver Y have shown that, once airborne, if they find themselves off course they can adjust their direction, powering themselves across the wind to readjust their bearing. Even with these navigational skills, how far they get and when and precisely where they make landfall still depends to some extent on the weather. Fortunately for eager migrant moth

Silver Y

seekers, skilful scrutiny of weather charts eliminates much of the guesswork.

By studying meteorological pressure systems and air-flows across continental Europe and North Africa, it is possible to work out which species will be on the move, their likely landing spots in the UK and their expected time of arrival. When conditions look promising, suggestions, hopes and dreams are shared on moth social media and dedicated migrant moth hunters will be out, their light traps strategically dotting the coasts, fingers crossed for the ultimate prize of something new. My involvement is largely vicarious, as I admire from afar the outcomes of people's follies, when photographs of successfully intercepted moths are proudly shared the next morning.[3]

In the UK, the lion's share of migrant moths arrives in southern counties, but there are species that regularly turn up further north. Dark Sword-grass (*Agrotis ipsilon*), Scarce Bordered Straw (*Helicoverpa armigera*), Rusty-dot Pearl (*Udea ferrugalis*) and Rush Veneer (*Nomophila noctuella*) are among those that reach my coastal traps in southeast Scotland most years.

Occasionally, but not too frequently to spoil me, weather systems conspire to carry an even greater variety this way. In July 2019, the pattern of pressure systems circling over Europe meant Scotland enjoyed more than its usual dose of migrant moth activity. Top of my list was a magnificent moth: Bedstraw Hawk-moth (*Hyles gallii*).

Just a few days before, at a moth-trapping session I run, a friend proudly paraded in front of me a Bedstraw Hawk-moth that had come to his garden trap. He had prepared me for the event by a crack-of-dawn email simply stating 'you'll never guess what I caught' in the header. He was right, even though my mind had rifled through a number of exotic possibilities. This superb moth was the first '*gallii*' to be recorded in East Lothian for at least a hundred years. I admired its chunky cigar-shaped body and smart dark olive-green wings, each traversed with paler stripes and streaks, and splashes of red. We carefully posed it on a tree trunk in order to etch its image onto our cameras' memory cards. Seeing somebody else's moth is never quite as good as finding your own, but despite the teasing I received on the greenness of my skin, I was more than happy to enjoy this one, pleased to have an opportunity to see one of these magnificent moths in Scotland.

Weather forecasts indicated that conditions over the subsequent days would be good for moths in general and migrants in particular, so a couple of nights later I rashly set up two moth traps in my garden. Close to my house and the mains power socket it requires was my biggest trap, lit with my all-powerful mercury vapour bulb. Although he made it, this is the trap my husband despises as its bright beam infiltrates through our bedroom window and disrupts his sleep. I compromise a little by positioning the trap in a less than ideal position to the side of the house and he does his bit by sleeping with a pillow over his face. About thirty metres beyond our garden boundary in a weedy stretch of field margin I put my smallest trap, lit with a much weaker battery-powered fluorescent bulb.

Next morning, I awake at 4 a.m., just as the sky is starting to lighten. Slipping silently out of bed, I creep downstairs, step barefooted into wellies and join the refreshing morning air outside. The large trap by the house shows it has caught lots of moths, but nothing around the edge immediately stands out as unusual to my sleep-hazy eyes. I stumble my way along the edge of the field to the smaller trap. Its bulb is still shining in the pre-dawn gloom and I can see there is something big resting on the plastic lid. As I get close its identity becomes obvious; it is my very own and very beautiful Bedstraw Hawk-moth. Not for the first time, I am glad I made a pre-dawn check. Within an hour the sun would have risen above the horizon, the moth probably flown and I would never have known.

Over the coming days, reports of other Bedstraw Hawk-moth sightings come in from across Scotland. Most are spotted in the eastern counties, but I hear of one from the middle of Scotland, near Aviemore. Another trapper has an astounding three in his small trap on Mull, and somebody else had an enviable experience during the day, watching one feeding from machair flowers on the spectacular coast north of Ullapool. There were others reported and I can only imagine how many more Bedstraw Hawk-moths there must have been in our night time skies that week, flying unseen, chancing it ever north-wards and westwards to an unknown destination.

Not all long-distance migrating moths are large, powerful fliers. Less than a centimetre long, the Diamond-back Moth (Plutella xylostella) makes up what it lacks in size by being extraordinarily abundant.

Diamond-back Moths have been dubbed the most wide-spread moth in the world. They live almost everywhere, even as far north as Svalbard, though in colder regions they are unable to survive winters and must repopulate entirely by migration each year. One of their less impressive accolades is 'super pest'. Their small green caterpillars devour the leaves of brassica crops such as cabbages, swedes and oilseed rape. Each female can lay several hundred eggs which develop into reproducing adults just a few weeks later. The warmer the weather the faster they develop and damaging outbreaks can happen quickly, especially in warmer sub-tropical regions. To make matters worse, they are resistant to many pesticides.

In the UK in early summer, if the wind is favourable, many thousands of Diamond-back moths might turn up on our coasts overnight, wafted across the North Sea from continental Europe. A few years ago in southern Scotland, a mass arrival made my mundane dog walks in the fields around my house slightly more exciting for a few days. As I brushed past oilseed rape overhanging the path, clouds of tiny moths took to the air. Crouching down among the crop, I watched them busy among the leaves, their silvery grey wings gently glistening and the paler zigzag markings (somewhat underwhelming diamonds) down their backs giving their identity away.

The timing and abundance of Diamond-back moths arriving in fields varies from year to year, but within a week, caterpillars will be hatching and getting to work on the leaves. Within a few more weeks, the next generation of moths is under way, powering onward on tiny wings. The timing of potential outbreaks can be predicted by farmers and researchers looking out for signs of caterpillars and adults and sharing the information with each other.

Recently an alternative forecasting system has been trialled by the Warwick Crop Centre at the University of Warwick,[4] using information gleaned from moth-trapping citizen scientists across Europe. There are plenty of moth recorders who regularly share moth finds with friends and colleagues on social media or upload records to biodiversity monitoring websites. Here is a daily supply of data already being provided, from numerous locations across northern Europe. When numbers start to pick up in enthusiasts' moth traps, they will also be building up in nearby crop fields. The data is easily collated on a website anyone can access, providing a far more complete picture and consequently a better advance warning system of moth outbreaks than local farmer networks alone can do. The end result is more effective pest management in particular by avoiding indiscriminate pesticide applications.

As autumn approaches, the falling temperatures and shortening day length trigger the British-bred generation of some migrating moth species to uproot. Unable to survive the cooler winter, their best chance of survival is to head south towards the Mediterranean to the regions their parents originated from all those months ago. Here, in relative warmth, they can reproduce and a new generation of adults will be ready to migrate northwards some time the following spring when the conditions are right. We now know that many of our familiar and regular migrants complete similar annual back-and-forth journeys over successive generations.

But there are other 'migrating' moths, perhaps better described as vagrants, for which the long journey north is a dead end for them and their progeny. For example, if the Bedstraw Hawk-moth that had arrived in my garden had laid eggs, even in the sultriest of Scottish summers there wouldn't have been enough time for the caterpillars to grow, pupate and develop into a winged adult that could return south before the onset of winter, when it becomes too cold for any of these life stages to survive. Why do thousands of moths bother to make such a fateful journey?

They do it because there is always a chance that some, sometimes, might end up in a location which is suitable for survival and a new population becomes established. These pioneers are important, ready to take advantage of the conditions in which they find themselves. Some species are regular vagrants, the ones that reach us in northern Britain perhaps the more optimistic individuals of an annual cohort that mostly stops much further south. Others are more unusual and unpredictable wanderers that are only very rarely found this far north. For moth enthusiasts eager to see as many different species as possible in Britain, it is these moths that get heart rates pulsing fastest.

When I was a birdwatching teenager living in Surrey, a golden-winged warbler turned up in a supermarket car park somewhere in the neighbouring county of Kent. I'm not sure how I heard of the news, for this was before the days of social media, but I have vague memories of a phone number you could call which would provide a recorded daily update of rarities. My mother dutifully drove me there. As she stayed in the car reading her book, I tried to join the throng of khaki-clad men equipped with telescopes and long camera lenses. I hated it. I quickly realised I didn't want to be in a tarmacked car park among a crowd of strangers to see my wildlife. I wasn't even sure I wanted to ogle some poor bird which had found itself accidentally off course in an urban corner of Britain. Surely it was tired and would want some privacy to rest its wings and refuel? In the end I never saw the warbler; I 'dipped', in twitcher parlance. In fact I don't think anyone saw it that day; probably the bird had sensibly moved elsewhere. We drove home. A wasted journey. Although my mother had the good sense not to say as much.

Twitching moths is easier. An unusual find can be kept in a pot in the cool and dark of a fridge for a day or two without harm, allowing time for others to drop by for a social visit and see it. I've not yet been presented with an opportunity to 'fridge-tick' a moth so I can't be sure what I would do. Would I travel to see somebody else's find? How far would I be prepared to go? Would it have to be a once-in-a-lifetime rarity or would any large and flamboyant moth tempt me? I'm not particularly interested in keeping a tally of the number of moths I've seen, but I do know that for any species, common or rare, resident or migrant, until I discover one for myself or catch it in my own trap, I won't add it to my list.

Perhaps the most surprising vagrant moth to have ever turned up in Scotland was a Bedrule Brocade (*Mniotype solieri*), found in a light trap in the Borders village of Bedrule in 1976. It is a far from flamboyant species, but at the present time this is the only one ever recorded in Britain. As a resident in the Mediterranean region it had strayed some distance from its usual haunts, but the consensus is that it probably arrived under its own steam. So exciting was the find that it was bestowed with a common name the same as the village in which it was found. Its memory lives on locally too; a narrow strip of grassland near where it turned up has become an unofficial nature reserve: the Bedrule Brocade Moth Wildlife Reserve.

One of the joys of light trapping for moths is the element of the unexpected. Some people liken opening a light trap to the unwrapping of presents on Christmas day – you never know quite what you will get. Setting a trap in suitable weather should catch some moths, and it's always worthwhile to enjoy the more predictable regulars. Migrants, on the other hand, are more of a gamble. Choosing a good location when the weather conditions are right will improve the odds, but a large slice of serendipity is also involved. Perhaps it is this mix of possibility and uncertainty that can make moth trapping so addictive. Success happens just about often enough to fuel the desire to try 'one more time'.

As autumn slides into winter, the potential for migrant moth excitement definitely dwindles and the addicted can hang up their moth traps and relax for a few months. Winter is a quieter, more sedate time of year for moths. Even so, for restless moth seekers, adventure still awaits.

CHAPTER 12

Coping with Cold

*I wonder if the snow loves the trees and fields, that it kisses them
so gently? And then it covers them up snug, you know, with a
white quilt; and perhaps it says, 'Go to sleep, darlings, till the
summer comes again.'*

Through the Looking-Glass, Lewis Carroll

Darkness comes early in November, allowing nocturnal moth
hunting to start in the late afternoon. My daughter's after-
school activity in a nearby village is an inconvenience for
most of the year. The half-hour slot doesn't leave time for
much more than a short dog walk, or if I'm feeling lazy, a
chance to sit in the car and read a book or inanely flick
through social media. However, once the clocks have
retreated an hour, this changes. It is suddenly dark, and with
it I have an opportunity to seek out moths.

In Scotland, the aptly named Winter Moth (*Operophtera
brumata*) and Northern Winter Moth (*Operophtera fagata*) are
the most abundant moths on the wing in November and

December. Taking advantage of predator-free skies, males are able to fly even on nights when the temperature is around freezing; these are the pale fluttering shapes that are illuminated in the headlights of your car as you drive home from work. The females' wings are reduced to non-functional stubs, so they can only get about by walking.

An adult Winter Moth's life is entirely about reproduction. The caterpillars pupate in the woodland floor in summer, just below the soil surface. As winter darkness falls, newly emerged moths push their way to the surface and crawl up the nearest tree trunk or convenient vertical surface. Here females summon potential mates by wafting their sex pheromones into the chilly air, and wait. Males in the vicinity are quick to detect these chemicals, homing in on the female to mate with her, hopefully before another male gets the chance. It is this tree-trunk tryst that I want to see.

Daughter dropped off (slightly early), I hurry towards the woods, breath smoking from my nostrils. A tawny owl t-woos somewhere in the distance and I pass a dog with a flashing LED collar leading its owner urgently homeward for dinner. As I set foot on the broad sandy path that leads through the woods, my torchlight immediately catches pale-coloured moths delicately fluttering between the trees. Sweeping the beam over smooth sycamore trunks, I pick out scores more; delicate pale grey shapes, peppering the trunks, waiting, for something. These are male Winter Moths. Individually they are unremarkable to look at – weakly patterned in shades of palest grey – but as my torch picks them out from the dark in numbers, they make quite a spectacle. I suspect these are newly emerged, not long crawled out of their subterranean pupa, getting used to their new wings and making ready for the night ahead. I have clearly timed my visit well; they are everywhere. On most of the tree trunks or weakly flapping through the air; probably the most numerous moth I have seen all year.

I make my way slowly forward, sweeping my torch carefully over every trunk that flanks the path. It is several minutes and well over a hundred male moths later before I finally spot something a little bit different and a little bit more exciting: a female. She is nothing like the male, unrecognisable as the same species. Her body is dark and plump, her legs spindly and she sprouts two tiny shoulder pads for wings. At first glance she looks more like a spider or fly as she clings to the trunk. Even at closer quarters, she doesn't really look like a moth. The male moths perhaps agree. Despite a clear surplus of suitors around, none have shown her any interest. Maybe she isn't ready for them yet, or perhaps she has already mated, despite the early hour.

I wonder if females really are so much less numerous than the males, or if I am simply very bad at spotting them. I expect I have already passed a few, taking them for a small knot or bark fissure, but surely, I wouldn't miss hundreds? Then again, what would be the advantage of having so many more males? A more likely explanation is it's still early for the ladies. In many moth species, males tend to emerge a few days before females. This means that males are ready and waiting and a newly emerged female doesn't need to hang about for her beau. She can maximise her short life with the business of laying fertilised eggs.

On the next tree I find a couple, abdomens joined tip to tip in their mating embrace. The female faces upwards, her egg-laden body impressively rotund against the male's pale scrawniness. He hangs beneath her, facing the leaf-littered floor, his wings held shut above his body like a miniature sail, standing out in my torchlight and betraying both of them as he gently sways in the slight breeze.

Once mated, the female Winter Moth continues her upward crawl, laying her eggs in bark crevices and forks of branches. Although it may not be the most adventurous of adult lives, it works well. By giving up the capability to fly, a female is able to invest more of her valuable resources into making eggs. The job of dispersal is delegated to caterpillars, and the breeze.[1]

I think to check my phone; it's 6.30 p.m. already. I'm going to be late, as usual. I'm eager to see more, but tonight I must leave these Winter Moths to it. I allow myself one more speedy check over one more tree, just in case it hosts something new, before ruining the magic of the moment by switching into mother mode and running back to the car. My daughter is already there, waiting.

High latitude winter is not an easy time to be a moth. Temperatures are low, nectar hard to come by and food for caterpillars in short supply. As we've seen, some species migrate to somewhere warmer, but the majority stay put, sitting out the coldest months of the year in a quiescent state known as diapause. Life is put on hold as they await the return of warmth, a new growth of plants and better opportunities for eating and moving about.

Some pass the time overwintering as eggs or pupae. Eggs are laid in cracks of bark or hidden in leaf buds for protection from predators and the coldest temperatures. Pupae are similarly hidden, for example under bark, in the soil or among leaf litter.

Caterpillars that feed on roots or on woody tissue deep within tree trunks are buffered from the winter conditions going on outside and may be able to continue business as usual, albeit a bit more slowly at lower temperatures. Those in more exposed situations simply become inactive when the temperatures are low, clinging to twigs or bark, squeezed into cracks and crevices or curled up in leaf litter or thick vegetation. For added protection, particularly if there is any chance that temperatures will plummet below freezing, they prepare their bodies by becoming more cold-hardy.

Anyone who has ever mistakenly put lettuce in the freezer will know the damage freezing does to cells. This is because water inside cells expands when frozen, rupturing membranes and turning everything to a mushy pulp. The only way to prevent such damage is to avoid freezing. Animals and plants can do this by filling their cells with chemicals which lower their freezing point; much the same as adding antifreeze to a car's windscreen wash tank. The precise cocktail of antifreeze chemicals, along with a few other cellular adaptions, dictates how much cold the cells can cope with, and for how long.

For caterpillars, the most extreme cold conditions, as low

as minus 70°C, are endured by the Arctic Woolly Bear moth (*Gynaephora groenlandica*). Named for its very hirsute and handsome caterpillar, as an adult this moth has wings beautifully patterned in various dark shades of grey, ideal for concealment on the rocky ground and soaking up rays of the Arctic sun. It is the world's most northerly breeding species of moth, eking out a remarkable life in the icy realms of Canada and Greenland. At these high latitudes temperatures only become warm enough for activity on sunny afternoons in midsummer, so it takes on average seven years, a severely punctuated seven years, for the caterpillar to complete its development.

During each brief summer, which lasts little more than a month, the caterpillar enjoys feeding, basking in the sun and growing. Then, in July, it prepares for winter by spinning itself a flimsy silken shelter in among rocks where it remains, inactive and mostly frozen, for eleven months. Antifreeze chemicals synthesised in the caterpillar's cells prevent damage by the ice, but this is a freeze-tolerant luxury that only the caterpillar stage possesses. When the caterpillar emerges from its final winter freeze it has a race against time to pupate, emerge as an adult moth, quickly find a mate and lay eggs before the ice returns. The tiny caterpillars hatching out have a brief time for a nibble of some leaves before it is time to hunker down for their first eleven months of cold.

Our own Winter Moths are lightweights in comparison, but are one of the few species that have made a success of being reproducing adults in winter. They don't last the entire winter; their flight period is usually over by the end of the year when all that remains are their eggs, waiting for springtime to hatch. Some British moth species manage to pass the whole winter as adults, but these remain more or less inactive during this time, hiding away within the protection and insulation of leaf litter, tussocks and crevices. They re-emerge the following spring to lay their eggs.

Luckily for the moth enthusiast, a handful choose winter hideaways that are spacious enough for a human to explore. The adventure of seeking out these moths has become a highlight of my moth year, some suggest an obsessional niche, providing me with a welcome moth distraction through what could otherwise be a quiet time.

'I can always tell when you've had a bath,' my daughter claims, with what I like to hope is affection. 'You leave bits of moss and twig behind.'

Related statements such as 'You stink of cave again' or 'You don't need to come into the playground to pick us up' have punctuated their childhood and my motherhood, although if you believe my older sister's reminiscences, my bed was also a 'bit gritty'. Perhaps the affliction of bringing the essence of my outdoor ramblings inside is one I was born with.

Sometimes my clothes might smell a bit damp and musty, my elbows and knees streaked with mud, and more than once when waiting in the supermarket checkout queue I've smoothed back my hair only to discover a bit of entangled bramble. It turns out that moth recording isn't only about sunny days with a net, or warm still nights with a light. From October until the following March it doesn't matter what the weather is like or what time of day it is; to find moths all you need is a torch and the promise of a dark place. The damp interiors of caves, mines and culverts are my go-to haunts as I try to find out more about some of our loveliest moths, which spend months of their adult life hiding in the dark.

The Herald (*Scoliopteryx libatrix*) is a striking moth. It is about two centimetres long with deep russet-brown wings adorned with a generous smudge of fiery orange on each shoulder. Delicate lighter lines cross the wings towards their apex and the wing edges themselves are gently scalloped. The furry thorax is styled into a diminutive Mohican-like crest over the head, and to top it all their dandy legs are striped black and white. Heralds emerge from cocoons in late summer and build up energy reserves by feeding, particularly on ripe autumnal fruit like blackberries. To help with this food choice the sides of their proboscis have jagged edges towards the tip, ideal for breaking into soft fruit to suck up the sweet juices. Heralds aren't ready to reproduce until the following spring, so once they have fed sufficiently, they seek out a cool dark place to spend the winter months, protected from cold weather and hidden from predators.

As the days start to lengthen and temperatures rise the following spring, they emerge from hiding to mate and lay eggs on willow, poplars and aspen, the caterpillars' food-plants. Although a common and widespread moth, it is unusual to see more than one or two adults at a time, in a light trap or at a sugared post. And so, one November a few years ago, when I heard on the moth grapevine that somebody had counted almost fifty Herald moths in a ruined castle my interest was piqued. Not only was Herald a moth I had yet to see, but the prospect of this many in one place was too exciting an opportunity to miss.

The ruin in question is hidden away in a large woodland. Reaching it involved a pleasant autumnal walk, kicking my way through paths of bronze and gold leaves and criss-crossing a burn as it rushed noisily beneath old stone bridges. The castle would once have commanded a fine view across a flat arable plain and the Firth of Forth, to the north and the rolling hills of the Lammermuirs to the south, but now it is shrouded by tall trees and gradually crumbling into obscurity.

As I crept into the gloomy entrance passage, it became apparent the beam of my small pocket torch was not quite as penetrating as I had hoped. But, by methodically scanning the ancient brickwork, it wasn't long before I started to pick out Heralds.

In the first section they were scattered in ones or twos but as I ventured down a spiralling flight of stone steps there were dozens lying together in random, overlapping associations on the ceiling above me. Through my anthropomorphic eyes it was as if they were snuggling up to each other, but more likely the arrangement was the outcome of a jostle for the perfect spot, or through seeking shelter in the lee of an earlier arriving inmate. Each one a beautiful moth, when huddled in the silent stillness of a ruined castle they made an amazing sight, worthy of a place on a list of wildlife spectacles.

Eventually, after counting and re-counting I made the total

seventy. But why so many in one place? Were they somehow being attracted to this dark castle from a wide area, picking up on a musty scent or detecting the smell of other gathering Heralds? Or was it simply that the Herald is a more numerous moth than light-trap records have us believe? I mulled over these questions as I sat down on the stone steps to tighten my bootlaces in readiness for my walk back. As I paused, I noticed another moth on the wall beside me. Silvery grey, with wings spread flat against the wall it was well camouflaged. It was a little battered and not as showy as the Heralds, but this was an even more exciting encounter, for I had no idea what it was.

Mark Cubitt is usually my first port of call for exciting moth enquiries so it was to his inbox that my email 'mystery moth' with accompanying photo was sent. And it was he who later confirmed that I had discovered Scotland's first overwintering record of a Tissue (*Triphosa dubitata*).

There had been a handful of Tissue recorded in Scotland before, but only sporadically in autumn or spring and always found resting out in the open by day, or in a light trap left overnight. The scarcity and randomness of sightings made it seem likely these were wanderers from further south, perhaps sometimes, briefly, securing temporary residency here. This overwintering one was a tantalising conjecture that maybe the moths were more firmly established. To prove it, I needed to see if I could find more. Mark and I decided to investigate further and, the following weekend, we set out on a mission to explore some dark places.

⋆　⋆　⋆

Beneath the surface of a wooded hillside in West Lothian are the abandoned passages of a small limestone mine. We follow an indistinct path, thick with fallen leaves, along the edge of the wood to a ruined cottage. Perhaps once a cosy residence, its doors and windows are now absent, the roof long gone and instead a brace of sycamore reaches through toward the sky. Entering the wood, we duck beneath low-reaching branches and clamber over moss-cushioned fallen trunks to our destination: a dark opening in the woodland floor and access to some of the tunnellings below. The entrance is small, giving two entry options: a feet-first or a head-first approach.

Unsure of what lay below, and on the basis that I would rarely consider descending a ladder head-first, I opt for the former. Lying prone with my feet poking into the hole and hands pushing against the leaf litter I reverse my way down. With my face centimetres from the ground, I can't help but breathe in the rich earthy smell, a wonderfully comforting concoction of moss, fallen leaves and rotting wood. Once inside, I find myself in a spacious chamber, with a smaller passageway disappearing into the darkness ahead. The temperature feels a little warmer than the November day I've left behind and the air is heavy with damp silence. Crouching, I cast my torch beam over my surroundings. A pair of size 13 wellies, rapidly followed by some legs, a body and a shower of leaf litter, soon intrudes on my peace and announces Mark's arrival from above. We begin exploring.

The walls of the passage rise a couple of metres high, sided by crooked layers of grey rock with alcoves, nooks and crannies concealing silent Heralds. Water droplets have condensed on their wings, bejewelling them in our torch beams. We advance slowly, navigating fallen slabs on the ground, trying to search the walls and ceilings methodically and silently counting the moths we see. After about twenty metres the passage narrows. As I stoop to fit myself through, Mark excitedly whispers from ahead, 'Tissue!' Lying on the stony wet ground, propped on an elbow, he shines his torch on the wall to reveal his discovery.

This one is much fresher-looking than the one in the castle ruin. It is a silky silvery grey, with a combination of delicate and bold black wavy lines traversing each wing. Rosy streaks add a subtle blush. Loveliest of all are the wings' undulating edges, outlined with a thin black line and fringed in delicate pink. I had discovered a new favourite moth, to my eyes more beautiful than the Heralds, and although I didn't know it then, a moth that has remained a firm favourite six years later. We lie back in the silent darkness, unable to see each other but well aware of our shared sense of joy and accomplishment. Overwintering Tissues in two places, eighty kilometres apart. Surely they had to be resident in southern Scotland.

In the days that follow, Mark and I exchange a flurry of emails; words and pictures are put together for social media and an online recording form created. Within two weeks our Hibernating Herald survey is rolled out, encouraging people across Scotland to look in their unheated sheds and garages, visit ruined castles and explore dark caves in search of Herald and Tissue. As records of Herald start coming in, we single-mindedly spend our weekends discovering and exploring as many dark spaces as possible. A huge subterranean world is revealed, along with an associated following of people – from speleologists and bat ecologists to urban explorers – obsessed with finding, accessing and exploring this hidden world of darkness. We lepidopterists are not quite in their league. The moths tend not to venture far inside, most finding sufficient darkness within fifty metres of the entrance. Deeper, more intrepid explorations can be left for others' fancies.

In the first five years of the Hibernating Herald survey, torch-wielding people across Scotland counted over 13,000 Herald moths from 360 sites. This is about forty times the number found during the same period in light traps or at sugar. For Tissues the results have been just as astounding with over 270 individuals recorded (before the survey began there had been just thirty Tissue recorded in Scotland, ever).

The challenge of discovering some of the harder-to-reach hideouts has been enjoyable, and our quest for hibernating moths has taken Mark and me on expeditions across southern Scotland. We've scaled precipitous cliffs, traversing ledges (in wellies) I'd rather not remember; been sleet-soaked on wind-blasted moorlands; had our skin punctured by battles with spiky gorse thickets and spent many frustrating hours going round in circles, searching for promised mine entrances that in the end turn out to be blocked off.

Sometimes we explore together, sometimes alone. Occasionally we coerce others to join our folly. My husband likes the walking part, but when damp, muddy crawling is involved, he opts to chivalrously guard the entrance, leaving me to enter on hands and knees and enjoy the moths and murk in solitude.

As each heralding season comes and goes, my knowledge of Heralds, Tissues and some of the other creatures that can be found in the dark culverts, caverns and ruins of Scotland grows.

Not all Heralds last the winter. Spiders cunningly construct their webs to intercept moths as they enter or leave. Those resting in dim half-light near the entrances risk discovery by winter-hungry birds. I once left a trail camera in a ruined building and, over the course of a few days, a wren flew in regularly and with purpose, methodically picking the Heralds off the walls until none remained. Each year new Heralds return to this ruin, each year a wren comes along and polishes them off. It's frustrating to see moths repeating the mistakes of the generation that went before, but enough adult moths must last the winter in nearby overwintering spots to maintain the population. Wrens need to survive too.

In Yorkshire, a small team of enthusiasts has been monitoring Tissues and Heralds in water-sculpted limestone caves of the Dales for nearly twenty years; far longer than our Scottish effort. Their surveys were started by David Hodgson, a keen caver throughout his life with many an entertaining anecdote of his youthful exploits underground, starting way back in the late 1940s. In his slightly more responsible but no less intrepid older years, he began documenting the wildlife he discovered on his underworld adventures more carefully.

Tissue Moth

David was puzzled why one cave in particular always hosted much higher numbers of Tissues, usually at least a hundred, and one year just over four hundred. Nowhere else in Britain have anywhere near such large numbers been recorded. With characteristic thoroughness, for the next decade he made visits to this cave by day and night, measuring and monitoring. Bats flying in and out were logged, water levels assessed, radon levels investigated, direction of draughts noted and noise levels recorded. Yet the reason this one cave attracts most of the valley's Tissues each year is still unclear. Are the conditions inside superior, does it smell better, or is it chance? Perhaps this is simply the cave that's closest to where the moths breed.

Joining the exploration were local naturalists Paul Millard

and Nyree and David Fearnley. One October half-term holiday, which I strategically and entirely intentionally booked in Yorkshire, I went to meet the Yorkshire team and see their Tissue spectacle for myself. Leaving the family to sightsee in York, I made the pilgrimage to the Dales for an expedition I had looked forward to for a long time.

We hadn't ventured far along the passage of the first cave before clusters of delicate Tissue moths started to pepper the pale cave walls, their slate-silvery wings catching in our torchlight. Most were still, but there was occasional movement as one walked up a wall, wings held aloft, to find a better position. So many Tissue moths together; this was a different order of magnitude to anything I'd seen just a few hundred kilometres north in Scotland.

Paul tells me how one year he reared around a hundred Tissue caterpillars through the summer. He and Nyree then spent a day painstakingly marking the wings of the new adults before releasing them into the valley. Extraordinarily one was recovered almost a month later in one of the caves, seven kilometres away. Did it fly there with purpose, or end up there by chance? Impossible to know, but Paul and Nyree have studies under way to investigate further.

Tramping the hillsides between cave entrances, Nyree and I of course chat about moths. She shares her experiences rearing Emperor Moths from eggs, I hear of her exploits lugging light traps up steep slopes and tales of moth-ing adventures in the Scottish highlands and islands. As we crawl the muddy cave passages and navigate a precarious ladder behind a tumbling waterfall, I recognise a kindred spirit; somebody else up for adventure in pursuit of moths, simply because it's enjoyable and she's curious to know more. 'Some people call us extreme moth-ers,' she admits with a grin. Exactly the same epithet Mark and I have been awarded on account of our own underground exploits in Scotland.

CHAPTER 13

Then and Now

plus ça change, plus c'est la même chose

Jean-Baptiste Alphonse Karr

I've sloped off for the afternoon to a small valley in the Lammermuir Hills. With barely a cloud in the lazuline sky and the wind only just managing to fidget the leaves, it feels perfect for a moth hunt. Then again, it needs to be; I've decided to mount a search for Heath Rivulet (*Perizoma minorata*), a scarce day-flying moth.

Heath Rivulet is small – it would barely conceal my fingernail – with delicate pale wings patterned in grey. It lives in grassy moorland habitats where the caterpillars feed on the developing seeds of eyebright, a small plant with pretty, tiny white flowers. It doesn't look like a moth that wants to get noticed, easily disappearing into anonymity by resting on wing-matching rocks or hiding in vegetation. I'm told the best chance of spotting one is when it flies. This is most likely when the weather is warm and calm; not a reliable combination of conditions in their upland homes, even in late July. Perhaps it's no wonder they are not often seen.

Today however the conditions are good, and even better, I know Heath Rivulet has been seen in this place once before, by an accomplished local entomologist named Alice Balfour. Not only is Alice's East Lothian moth list enviably long, with several species I have yet to see, but she has a canny knack of beating me to it. So often when I've identified something unusual, something I think might be a new moth species for the area, it turns out Alice has already found it here.

I stroll along the valley, sheep eyeing me warily, scattering in panic long before I get close. Perhaps they're not used to lone walkers or maybe the white bag of my net makes them uneasy. They have grazed the grass short and swathes of bracken are encroaching; it doesn't look particularly ideal Heath Rivulet territory to me, but having never seen a living one, I'm hardly an expert.

As I brush through the bracken a pale moth takes to the air in front of me. My hand instinctively tightens its grip on my net and I take a swish at it, though already I know this won't be the moth I'm after. It's too big. As I carefully inspect my captive I am proved right. It is a Twin-spot Carpet (*Mesotype didymata*). Similar colours and definitely a moth, but that is as close as it gets to Heath Rivulet. I let it go.

I reach a steeply sloping grassy bank with a small amount of eyebright growing at the base. This looks a bit more promising, but there is no sign of moths, let alone any rare ones. I prod the flowers with the end of my net, in hope of ousting any occupants, but a few tiny flies are all that take to the air. I saunter on. The next hour is spent in a similar fashion. I explore rocky areas, poke at heather, waft my net over grass and rushes. I spend time by the stream and time up on the ridge. I stare with intent at all the eyebright I encounter. More Twin-spot Carpet, a Green Carpet (*Colostygia pectinataria*) and several Common Carpets (*Epirrhoe alternata*) are swept up in my net, and each time I'm left wanting something else.

Finally giving in, I find a bank of heather to sink into and relax to enjoy some chocolate and the scenery. It looks like Alice will remain one up on me for this moth. Maybe Heath Rivulet don't live here anymore, perhaps the sheep and bracken have changed it to something they don't like. After all, when Alice recorded Heath Rivulet in this part of Scotland, it was 1913; a little over a century ago.

Alice Blanche Balfour (1850–1936) grew up with a love of natural history, roaming the countryside around her wealthy family's estate in East Lothian with her siblings, catching, curating and documenting what they caught.

Her eldest brother Arthur duly inherited the estate. He was a bachelor, and once Alice's elder two sisters married, the responsibility of managing the estate and Arthur's affairs passed to still-single Alice.

Alice took her domestic role seriously, ruling the household with capable and at times annoying stubbornness, only occasionally complaining that she never had enough time for herself. She was well known locally for her interest in butterflies and moths and made valuable contributions to the research of other zoologists, but such were her duties in supporting her brother's high-profile political career, including a stint as Prime Minister, that for several decades she devoted little time to her own entomological interests.

When she was sixty, her friend the eminent chemist Professor Raphael Meldola came up to Scotland and they went moth hunting together. The visit left her inspired and, with diminishing commitments as her brother's career slowed, the next fifteen years became the most productive period of moth recording in Alice's life. I was reassured to discover some of her most significant finds happened during her sixties and seventies; maybe my best moths are also still to come.

Alice bequeathed her impressive collection of pinned specimens, notebooks and equipment to the National Museum of Scotland in Edinburgh. Fast-forward eighty-five years and now I, a present-day moth recorder, can visit the museum's collection centre to admire her moths and pore over her notebooks and catalogues. Thanks to this well-curated legacy I'm able to discover more of Alice's world and the moths she once encountered in the very same places I frequent with my own traps and net.

In her notebooks, the inked curls of her script list details of many moths she saw, not just those that ended up on her pins. She notes their abundance, what they were doing, what the weather was like. Occasionally she marvels at her triumph over finding something unusual, but more often space is given over to recounting disappointment.

Alice Balfour's specimens

The weather has been beautifully fine and sunny, and we had a good many warm evenings, but one can't have one's guests and go off mothing!

Moth potted, but escaped just as lid was being put on.

Sugared on night of 16th. Warm still night. Not one moth came.

Reports of wildlife observations are so often biased to recount only successes. Poor or disappointing results are rarely published in journals or paraded on social media and must frequently be lost to history. It was refreshing to see that this lady whom I had long admired for her evident mothing skill not only sometimes suffered the same frustrations and failures as I do, but was also willing to admit it. Her evident modesty also makes her more unusual finds more believable.

As I leaf through her notes and read accounts from her family, I create my own picture of Alice. She was part of an influential household and an able and accomplished naturalist but, through circumstance as well as choice, spent much of her adult life tied up with politics and duty. Correspondence shows she was respected in academic circles. Her sharp mind questioned what she found; she had the motivation to take to the hills with her net and the dedication to puzzle over the identity of tiny specimens in her boudoir. Her know-ledge of East Lothian's moths was, and perhaps still is, unrivalled. I would like to have met her.

Like other naturalists, including Alice, I am just a tiny part of a long history of moth recording in Britain. My own finds offer a brief snapshot of the moths of here and now, but to make sense of their significance, they need to be considered in rela-tion to what went before (and if I could, what's to come after). As we saw in Chapter 4, analysis of comprehensive data collected by moth recorders across Britain in the past fifty years shows countrywide trends of moth comings and goings and, in general, a worrying decline in their abundance. If I returned to the places Alice once caught moths, would I, a hundred years later, find much had changed in this corner of Scotland?

With permission from her great-great-nephew, each month for a year I distributed my traps in the woodlands and parkland of Alice's former home. After each overnight session

I tallied the moths and compared them with Alice's records. It's hardly a fair competition when your opponent is no longer there, but I felt a sense of accomplishment when I found the same species she had and frustration when something I ought to have seen evaded me.

Some species that eluded me have almost certainly disappeared from the area. Alice used to catch Red Carpet (*Xanthorhoe decoloraria*) as it fluttered in late-afternoon sunshine along the grassy valley of the Whittingehame Water. Now this moth is increasingly restricted to higher altitudes in Scotland and only once reported in East Lothian in the last twenty years. Brindled Ochre (*Dasypolia templi*) came regularly to Alice's light traps, but despite making some effort to seek it out in likely rocky locations, it remains a moth I have yet to see. Perhaps I never will. Numbers of this beautiful tan-coloured moth have declined drastically in many parts of the UK; our increasingly warmer, wetter winters mooted as a possible reason.

Other moths have spent the last century expanding their range northward; species Alice could barely have imagined finding in Scotland in her lifetime. Given the care she spent nurturing her gardens, she might be pleased that Maple Pug (*Eupithecia inturbata*), a small, delicately patterned moth, is now living among the maples she planted. Another newcomer is Blair's Shoulder-knot (*Lithophane leautieri*). This smart moth of autumn, with wings streaked in various shades of grey, resembles a short fragment of wood, providing perfect camouflage on textured grey bark. It wasn't found in Britain until 1951, on the Isle of Wight. Alice had probably never heard of it. Now it is increasingly widespread in southern Scotland and making its way ever northwards as it takes advantage of non-native cypresses, including that ubiquitous favourite of garden boundaries and neighbourly disputes, leylandii.

Modern-day moth trappers with their battery-powered traps have it easy. I can take a bright light almost wherever I want and leave it shining to catch moths while I sleep. Although Alice had a light trap, it was only used with mains electricity at her house and the bulb would have been dimmer than the bulbs I now use. As far as I can tell, she seldom risked leaving it running all night. Instead, she spent many more hours than I do searching for caterpillars, painting tree trunks with a sugary mix and wandering around with a lamp at dusk to discover her moths.

As I creep around the boundary of the estate's now empty walled garden in the failing light, I enjoy imagining the redoubtable Alice doing the very same on a similar evening a century earlier. She notes an Early Thorn (*Selenia dentaria*) caterpillar on her apricots and an Angle Shades (*Phlogophora meticulosa*) 'hibernating in the vinery', conjuring up images of a way of life very different to the estate of today. She reports the lime flowers 'abounding' with Common Wainscot (*Mythimna pallens*) and 'hundreds' of Silver-ground Carpet (*Xanthorhoe montanata*) flying over the grass. Was she exaggerating, or does this hint at a greater abundance of moths back then?

I may not be as practised at dusk prowling as Alice once was, but my evening tally is best described in ones and twos. Even with some optimistic overestimating I would struggle to top 'several' as a description of my numbers. It seems likely these species are less numerous flying among vegetation than they once were. On the other hand, the variety and number of moths caught by the brightness of my modern light traps might have wowed Alice; her notebook suggests her trap caught far fewer.

Comparing actual moth numbers from one century to the next is hard, not only because the methods used influence the species and the numbers seen, but also because our interpretation of abundance is so often made in relation to what we are accustomed to seeing. Alice's quantification of 'many' could be very different to mine. This is one reason why submitting our moth observations into national recording schemes is so important. Spreadsheets of numbers and locations might not be much to look at, but analysed with care they can take account of recording bias and hazy memories, to reveal meaningful trends.

My early days of garden moth trapping are still memorable. The range of colours, shapes and patterns in a summer moth trap was both exciting and daunting. Some moths are distinctive and, I found, relatively easy to identify and remember, but what about all the rest? Where on earth was I to begin with these? My tactic was to photograph everything and retain a few of the trickiest in the cool of the fridge, secured in small pots, to mull over during the day.

Alice in the late nineteenth century used large, multi-volume moth books to help with identification, their sparse illustrations so different from the modern webpages and guidebooks I use, where just as many pages are devoted to detailed colour images as to text. Some moths remain tricky no matter what resources you have, and Alice's notebook scribbles, crossings-out and question marks show she struggled with identifying the very same groups of moths I struggle with now.

Once I have exhausted my library of moth books and webpages, if I am still stuck on an identity or feel in need of a second opinion, I can email a digital photograph to an expert. Many British species can be identified confidently from a photograph, and depending on whom I ask, I may even get an answer the very same day.

Alice wasn't without friends and colleagues she could call on for help, but without digital cameras or the internet, seeking a second opinion was far more laborious a hundred years ago. It must have been even harder for budding entomologists without the connections Alice enjoyed. 'Goodness knows how they managed to identify anything back then!' remarks one of Scotland's current micro-moth experts to me.

Loose bits of paper tucked between pages of her notebook reveal lists of pinned specimens she organised to send down to experts in Oxford or London. Some of her moths waited several years before she finally got to put a name to them. It must have been much harder, or at least slower, to learn from your mistakes back then.

For some, seeing a dead moth on a pin equates to the decline of a species in the hands of humans; surely if naturalists hadn't taken specimens decades ago there would be many more moths around today. But the number of moths taken by lepidopterists is a minuscule fraction of the whole, and for most species their short natural lifespans and high reproductive rates allow them to cope with these additional fatalities. Specimens carefully curated and documented in museum collections are now an invaluable and irreplaceable resource for scientists. It is only through the well-curated legacy of Alice Balfour that I can get an idea of the moths that once lived in the places I now roam. It is only because of her specimens that, using more modern knowledge, some of her original identifications have been confirmed or corrected. Specimens can inform us about the world's biodiversity both now and in the past, and the carefully considered collecting of them is still important.

Keeping a physical specimen has many uses. Microscopic examination is necessary to identify some species with certainty, particularly in less well-documented regions of the world. Any part of it can be measured, or even dissected, at any time in the future, and DNA can be extracted and sequenced. This not only helps with identification and taxonomy, but provides valuable information on how different populations are connected and how they have changed over time. DNA sequencing is now a key tool in wildlife conservation and ecological research. As this and other molecular techniques improve, they will continue to aid our understanding of the natural world and provide additional ways to go about curating, documenting and protecting biodiversity.[1]

For the amateur naturalist though, it is advances in computer technology that have probably most revolutionised wildlife recording as a hobby. Electronic communication between moth recorders around the world is now easy and quick, and we can store, share and analyse vast amounts of data about the moths we see, helping to inform conservation efforts worldwide. Computers are also having an increasingly useful role in helping people identify what they find. Within the decade or so that I have been recording moths with purpose, handy mobile apps have become widely available offering identification assistance from your pocket, wherever you are.

I recently joined a volunteer group to help scythe some grassland on a nature reserve. As we enjoyed coffee and cake to celebrate our stubble, someone spotted a hairy caterpillar idling past. I was called upon to offer an identity.

'Ruby Tiger,' I pronounced with some confidence, but immediate self-doubt forced me to add, 'I think.'

'Why don't I check?' said one of our group, proceeding to photograph the caterpillar with her phone. 'I have an app,' she explained.

Within seconds she had an identification. 'Drinker moth,' came the answer.

This, I knew from its appearance, it wasn't. I also knew that this was not the right habitat for Drinker moth, and that it hadn't been recorded in East Lothian before. 'No, I don't think so . . .,' I ventured.

Scrolling down, she brought up another identification: Ruby Tiger. The accompanying photograph was a good match, and the additional information fitted. Second guess isn't perfect, but it isn't bad. We had an answer almost instantaneously and without needing to flick back and forth through hundreds of pictures in a guidebook.

For people new to moths, or those without the time or inclination to invest in the steep learning curve of identifying their finds, apps provide an easy way to put a name to a moth. In some apps, with a few more clicks and swipes, details of the observation can be submitted to a recording database.

Tom August is a technophilic ecologist; he looks on gizmos and gadgets as providing all sorts of exciting opportunities to make life easier both in his home life and at work. His research includes projects that use artificial intelligence, both to help more people record wildlife and make the resulting data as valuable as possible for science. Meeting up over Zoom, he gives me an enthusiastic low-down on how to teach computers to identify moths. But before he starts, he offers a word of warning: 'You do know that most of what I tell you will be out of date in a couple of years.' It is the end of 2021.

Moth identification apps are made by training a computer. This training, or machine learning, involves supplying a computer with hundreds of thousands of already identified images of different moth species in various positions and poses, and leaving it in an air-conditioned room to crunch its way through them, learning which shapes, spots, mottles and stripes are characteristic for each species. Even with powerful computers working non-stop this process can take weeks. The best training includes extra information, such as the location and time of year a species flies, to help it come up with sensible suggestions. This should prevent obvious mistakes such as identifying a picture of a moth found in Britain as an Australian species.[2]

Moths, with their patterned wings and distinct shapes, turn out to be one of the easiest insect groups for a computer to learn, but even so, there are plenty that the computer finds hard to recognise. Even moth experts with decades of experience sometimes struggle – could a computer ever be clever enough to provide reliable identifications for these tricky species? Could a computer ever be *better* than a human?

Tom admits that some moths, the very same species that cause identification difficulties for people, will probably always be out of the realm of automatic recognition apps. 'But that doesn't matter, as long as you have an honest robot, one that owns up to how certain it is of its identification,' he says. The problem comes when computers present their suggestions as the definitive answer, or when users accept the identification without question.

This problem is not restricted to robots, in my experience. There are plenty of over-confident humans too, who don't always get it right. The best computer apps guide a person towards a correct identification, rather than provide a single answer. They can list possible identifications and give an indication of which ones are the most likely. Just as with any advice, it's up to the user to consider the options given before reaching their final decision.

Robotic moth recording goes a stage further with automatic light traps. There are several approaches, but in essence, a camera records images of insects landing on a white sheet suspended below a bright moth-attracting bulb. A trained computer evaluates each arriving insect and also tracks them so they aren't counted more than once. Images that it decides are moths are sent to another computer, a computer trained in moth identification.

This technology offers many possibilities, from non-lethal moth surveys in remote areas, to recording pest species on farms, to monitoring moth activity throughout the night. There are of course limitations and the technology could never completely replace a trained entomologist, but it can free up that expertise for more demanding projects. 'It's a big area of research right now,' Tom enthuses, 'and there are loads of people around the world working on it.' He is one of them.

As full of potential as all this technology is, one thing it could never replace is the joy of being in a dew-laden meadow at dawn puzzling over the contents of a trap. Even techno-phile Tom agrees: 'Part of the pleasure of recording wildlife has to be the challenges of identification and the reward of learning new things. Most of the time people won't want to delegate that to a robot.'

When I was beginning moth trapping and facing a bewildering array of similar-looking insects at dawn, I wonder if having an identification app on my phone would have helped me. Was my learning experience better as a result of the painstaking, frustrating and at times fruitless flicking back and forth through pictures in my guidebooks? With my introverted ways, if a computer had replaced the need for me to contact actual humans for help, would I have become a part of the Scottish moth-er community so easily? Perhaps it wouldn't have made any difference how I learned the ropes, but some of those who provided such patient support as I was starting out have since become good friends. My children are forever telling me, with rolling eyes, that I don't 'get' modern technology; perhaps sometimes I don't need to.

The accuracy of information apps is improving rapidly and the scope of using AI technology for moth recording is huge. Although a primary aim of identification apps has to be to provide the correct name for a moth so that the record can be used by conservationists, they also have an important role to play in simply engaging people to look more closely at nature. Even if the details never make it into a national recording scheme, if an encounter with a moth and a mobile app inspires one more person to be curious about the wildlife around them, then the computer has done a good job. Whatever their current limitations, these apps can open up the traditionally fusty and dusty world of entomology to a new audience, who may then be inspired to learn more and take it further.

Sitting at her desk among pins, setting boards and hefty hardbacked tomes, Alice Balfour would have found the concept of a computer bizarre, let alone a device that identifies moths for you. To her, the brightness and portability of my light traps would be impressive, even the sturdiness of my walking boots and the protectiveness of my anorak no doubt improvements on her own kit. Would she be envious of my digital camera and colour guidebooks, or would these resources get cast aside as new-fangled luxuries she could manage perfectly well without?

Whatever our age difference, if Alice and I could crouch together, under the glare of a light bulb on a warm summer's evening with moths batting around us, I think we would find much in common. Tools and methods change; the moths do too, but the motivation of naturalists to seek, document, study and enjoy the wildlife that shares their countryside remains much as it was a hundred years ago. Whatever the techniques and technology of the future, I hope people continue to have a curiosity for local nature, to seek it out and take comfort from it.

CHAPTER 14

Enjoying Moths

Interest in natural history and green things
is a secret of a happy life.

Miriam Rothschild

'You'll never guess what I've seen,' I remark excitedly as the children troop in wearily from their first day of school after the summer break. They don't enquire further, but I tell them anyway.

'There are twelve Old Ladies in the culvert!' That stops them in their tracks towards the biscuit tin. They are momentarily confused by news of what they know deep down must surely be wildlife-related, but then again perhaps some mad old ladies of North Berwick really have taken up residence in that damp culvert that their mother visits from time to time to check for moths.

I show them the pictures and normality is restored. Some moths are living in a culvert; what's the news in that? I change the subject and ask instead about their day. There is little intelligible response, other than it had nothing to do with old ladies.

Old Lady (*Mormo maura*) was a species I had long coveted from the pages of my guidebook: not only is their common name intriguing, but the large dark illustration regularly caught my eye as I flicked through. On reading the text I discovered it was widespread but not so easily seen; a species that is unusual in light traps and much better encouraged by an elixir of treacle laced with alcohol. This was a moth I wanted to meet.

Finding them in a local culvert was a surprise. I was there on the off-chance of another moth, a quick diversion from a tiresome supermarket shop. As I sloshed my way into the musty darkness my torchlight picked out their dark silhouettes against the brickwork. Not just one, but twelve, and much better than my previous imaginings. The dark wings are decorated with a patchwork of shapes in the deepest browns and greys, with a silky lustre, and at their margins the hint of a frilly edge. Their name is said to derive from the dark shawls that old widows once traditionally wore. Some were resting alone on the rough walls, but five had piled together in overlapping layers in a corner. It was hard to make out where one moth ended and the next began. Outside, above me, I could hear the rumble of traffic. From the skatepark came the rattle of wheels on concrete ramps, children screaming, a dog barking. Yet in my dark confinement with the Old Ladies, I was in another world, their secret underworld. What were they doing here?

I tried to find out more information, but there wasn't much. Summertime gatherings in sheds and tunnels are well known and generally documented as aestivation, a hot-weather retreat akin to hibernation in winter, although to escape high temperatures and drought conditions. Fair enough in southern Europe, but surely, I thought, Scottish summers can't be too hot or dry?

Enquiring of other moth recorders, I discovered very few reports of aggregations of any size in the UK, and none yet from Scotland. Somebody shared a photo of a group of perhaps fifty individuals, piled upon one another in a graffiti-covered section of underpass somewhere in the middle of England. I was sent another impressive photograph of an even larger gathering from the Netherlands. I received several half-chuckled tales along the lines of, 'I once had an Old Lady in my bathroom!'. But whether they are rarely reported or rarely seen, records of more than a few individuals at a time appeared unusual.

For the remaining weeks of August I pay my Old Ladies near-daily visits. Marking their wings with a tiny dab of paint reveals nightly comings and goings. Some stay for a few days without moving position, then they are absent only to return a few nights later. Others go and never come back. Visiting just after dark, I sometimes catch them joined by abdomen tips, wings brushing up against each other, silently procreating. Returning just before dawn, I watch them fly around in the culvert in search of a perfect daytime resting space. If I sneak there in the middle of the night, it is quieter but a few individuals remain resting on the walls, having a night off. My son enjoys regular evening lifts to the skatepark, and when I emerge from my Old Lady encounters he is eager to show me his latest skateboarding antics. Perched on the edge of the cold concrete ramp, I watch as he perseveres to perfect new tricks with silly names.

I now look forward to an annual catch-up with my local Old Ladies. Their arrival in the culvert in the final weeks of July marks the start of my underground moth season; it's also the middle of the school holidays, the start of the busiest tourist time here and a reminder of a forthcoming birthday. My son no longer wants my audience at the skate ramps; it turned out that first summer of Old Lady-watching was the last summer that I was allowed to be seen admiring his tricks in public. Now I must stay away when he is there with his friends. If I don't, he ignores me. Maybe between him and his friends, I am just that weird woman with a torch.

I'm still not sure why Old Lady moths congregate as they do. It seems unlikely they are escaping summer heat in Scotland. Could it be that they need a period of time after emerging from their pupa to mature? Gathering in a safe, dark place for the latter weeks of July might allow them to size one another up and wait until they are ready to do all the important things that adult moths need to do. Perhaps it is a redundant behaviour retained from more tropical origins thousands of years ago.

When I have more time, I will delve into the world of my local Old Lady moths again, following their nightly antics to try to understand more about how they live their lives.

I enjoy having small-scale moth projects like this to dabble in and amuse me, so it's always interesting to hear of other

Old Lady moths

people delving into the behaviour or ecology of a moth that has sparked their curiosity. One summer I was delighted to be invited to spend a morning on a bog, helping fellow moth enthusiast Ellie Lawson with her research.

Flanders Moss is a vast expanse of squidgy, squelchy lowland raised bog stretching out beneath an even bigger expanse of sky. Having formed over 8,000 years ago, the Moss is celebrated for being both ancient and big, but at eight square kilometres it is a fragile remnant of a great mosaic of bogs that once covered a swathe of central Scotland. Even though it lies within striking distance of the city of Stirling, many people bypass it as they head north to the alluring, and more stereotypically Scottish, skylines of the Highlands. However, although the contours couldn't be more different from its popular craggy neighbours, the bog's wide-open vista beneath ever-changing skies offers an accessible sense of wilderness.

My visit was early in July. The quest: to help search for caterpillars of the Rannoch Brindled Beauty (*Lycia lapponaria*). Rannoch Brindled Beauty is an enigmatic moth that wasn't discovered in Britain until 1871. Although possibly scattered across the Scottish Highlands, it hasn't been found in many places, and as I tried to discover more about it, I realised its lifestyle is not particularly well known either.

Ellie Lawson is a conservation trainee with NatureScot, working mostly on bogs. She is also coming to the end of an MSc and it was the need for a research project that initiated her studies of Rannoch Brindled Beauty. Joining us were Bethia Pearson, another enthusiastic trainee, and the reserve manager of many years, Dave Pickett. I couldn't have been in better company for a caterpillar hunt or to learn more about this soggy landscape. As we made our way onto the bog, Ellie told me more about the moth and explained her plan for the morning.

Rannoch Brindled Beauty is not your average sort of moth. The females have no wings; they look more like small blobs of fluff than moths. This means, that after emerging from a cocoon which has nestled underground over winter, sometimes for several winters, they have little choice but to crawl to reach as lofty a position they can manage. On a bog with few bushes or trees, this is rarely more than a metre or so up. Fence posts or, more naturally, tree trunks and tops of heather tussocks are their usual choices. Here they advertise their presence to the local males by releasing pheromones.

The males are equipped with wings and can fly, but there is uncertainty about their aerial prowess. My guidebook states males 'presumably fly', and in a much older reference I found them described as 'decidedly sluggish and disinclined to fly.' Whether they don't fly, aren't attracted to light, or maybe traps are never operated in the right time or place to lure the them, it seems males, just like the females, are easiest to find in their daytime resting places.

'Have you ever thought to throw a male in the air to see if it flies?', I asked Ellie. Of course she had, although her thoughts for moth welfare had so far stopped her short of trying.

There are other mysteries concerning this moth. As adult females don't fly, they must lay their eggs within crawling distance from where they emerge and mate. The duty of longer-distance dispersal thus falls to the caterpillars which, soon after hatching, spin a short thread of silk to catch the wind and lift them aloft. At the mercy of wind strength and direction they can only hope to arrive somewhere with suitable food.

If we are to believe the books, the favoured foodplant of the caterpillars is bog myrtle, a smart yellow-flowered plant of soggy wet places, but heathers, bilberry and willow are also known to be palatable. It's hard to know if these preferences are studied fact or supposition. Perhaps the moth has different tastes in different places: they may eat other plants too. With dispersal to new areas being at the whim of the wind, they can't afford to be too choosy. Maybe the caterpillars are easier to spot on the smooth stems of bog myrtle. If a popular guidebook states that a particular plant is the place to look, that is where the majority of subsequent observations will come from. Ellie was interested to find out for herself what Flanders Moss Rannoch Brindled Beauty caterpillars eat and where they hang out. She is also looking to see if browsing and trampling by the oversized local red deer population has any impact on caterpillar survival. Caterpillar counting could be a good way not only to monitor how well the moth is doing on the reserve, but also to provide information relevant to its conservation and that of other bog specialists.

Our task that morning is to find as many caterpillars as we can. Ellie suggests we spread out and walk in parallel lines, scrutinising the heathers and bog myrtle at our feet as we go. Like crime scene investigators, between us we hope to cover the ground effectively and find our quarry. 'They are quite obvious,' Ellie assures us. 'Once you have your eye in.'

As we slowly squelch our way along our allotted tract of bog, Dave submits to my endless questions about Flanders Moss and bogs in general. Peat bogs are important habitats. They store huge quantities of carbon, far more than any forest, and the mossy landscape has capacity to soak up vast volumes of water, keeping flood water out of harm's way. They are also home to some of the country's rarest plants and invertebrates and provide a valued sense of wilderness to people wanting to escape their urban lives for a short while.

But British bogs are in a poor state. After centuries of trying to tame them for our profit, bogs have been damaged, many completely destroyed. Large-scale destruction of Flanders Moss started in the mid-eighteenth century, and over a third of its original expanse is now farmland. Early labourers cut away the peaty surface by hand. Some of it was used for fuel or for building houses, but on large clearance projects, without any better ideas, turves were dumped in the River Forth to float away to the sea. Of course, there were some enterprising folk downstream in Fife who hooked bits out, dried it, and took it home for fuel.

When chipping away at the peat lost its appeal, or rather ceased to be economic, attentions turned to using the land to make money from trees. Ditches were cut to drain out the wetness and conifer trees planted as a crop. As the bog dried, its special plants and animals became deprived of their water-logged home. Some never returned.

Conservationists first started to protect bogs in order to save their rare plants and animals. Remarkably, it wasn't until the mid-1990s that their value as a carbon store and in flood prevention were properly realised and it's only really since then that bogs have at last begun to get the full attention they deserve, with projects to make them soggy again now under way in some places. Much, much more needs to be done, not least to end all commercial peat extraction for horticulture, but for Flanders Moss, its rewetting appears to be going well. As the water table rises, a healthy variety of sphagnum mosses is starting to thrive and the slow process of peat formation resumes. It is an important refuge for many plants and invertebrates, not just moths.

Nearly an hour of trudging later, my neck was beginning to tire of bending downwards and my eyes and brain were working together to convert heather sprigs into imaginary caterpillars. I was beginning to wonder just how many real caterpillars I might have passed over. At least my more experienced comrades hadn't found any either. Then, at last, Ellie spots our first. Squelching over to take a look, it takes me a while to notice it, despite having a pointing finger to follow.

It is stretched ramrod straight along a heather sprig. The body is the same grey as the heather stalk, with darker flecking to improve the match. Along each flank is a broken line of pale yellow punctuated with darker spots. Ellie is right, it is obvious once I see it but I've little confidence that my last hour has been spent looking for quite the right thing.

Spirits lifted, and with a better 'search image' in mind we return to our positions and resume the trudge. Ellie soon finds another, then Bethia spots her first. The pressure builds for Dave and me; jokes that you aren't allowed home until you've found one are exchanged.

We do find caterpillars, just not the right sort. There are young Emperor moths (*Saturnia pavonia*), lime green adorned with raised hairy warts of yellow and black. We find plenty of Beautiful Yellow Underwing caterpillars (*Anarta myrtilli*), billiard-table green with black, pink and white chevrons perfectly matching the sprigs of heather on which they sit.

There is an adult too, wings a rich chestnut brown with spots and streaks of white; the startlingly yellow under-wings hidden beneath are never revealed to us. We find the impressive Drinker moth (Euthrix potatoria), a large peculiar-looking creature, with broad russet wings and a furry womble face. The name Drinker comes from the drinking habit of the caterpillars, which dip their heads into water droplets before lifting them up to swallow. The adult moths don't feed, or drink, at all.

Rannoch Brindled Beauty caterpillars are notable by their absence.

After a break for lunch, we return along a different line, still searching the ground beneath our feet. It isn't long before Dave finally finds his Rannoch Brindled Beauty cater-pillar, leaving me as the only one who has so far failed. I gamely laugh it off with an 'I don't mind, I'm enjoying it and I'm pleased to have been able to see some', but of course what I want more than anything right now is to find my own. I have a caterpillar-hunting reputation to maintain and a quietly competitive streak to assuage. Finally, thankfully, one makes itself known to me, munching in plain sight on a tasty sprig of heather. Relaxing at last, about ten soggy steps later, I spot another.

Four lookers, three hours and six Rannoch Brindled Beauty caterpillars. A rather paltry return for our efforts. Ellie marks the location of each find with a flagged bamboo cane and will return to record details of the plants and the dampness at each. In the weeks after my visit, she continues her quest for Rannoch Brindled Beauty caterpillars on other parts of the bog. She even went on a night-time sleuth, where a grand total of one caterpillar was spotted.

Her season's aim was fifty caterpillars; in the end she spotted a very close and very respectable forty-eight. With these came forty-eight bits of information about what they eat and where they hang out. The data is yet to be fully analysed but Ellie concludes they are a 'pernickety species', seeming to like it not too wet or too dry, the vegetation not too long or too short. As well as the widely reported heather and bog myrtle, she spotted caterpillars on willow, cotton grass and even one lounging on a cushion of moss. Given the dispersal methods of the caterpillar, perhaps they need to be catholic in their tastes. In a blog summarising the fieldwork Ellie writes;

> It's been an incredibly interesting project so far . . . Slowing down to look at bog-specialist plants has allowed the whole team to see new things that would be otherwise overlooked, even those who've worked on Flanders Moss for years!

She's not sure where her work will lead her next, but has no plans to abandon her attentions to Rannoch Brindled Beauty moths. The more we know about their habits and requirements, the more we can do to help them to thrive on the bog.

Moths offer great opportunities for study, to help monitor the environment, learn about natural history around you and in the process notice a host of other wildlife. For me, getting to know a single species, asking questions about its adults and caterpillars, then finding ways to try to answer them is far more rewarding than simply adding to a list of species.

But moth recording, or for that matter any wildlife interest, goes beyond the individual species and beyond the questions and answers. It provides an opportunity for escape from the treadmill of normal life. This escape need not be ambitious. Even in my own garden, at 5 a.m. on a dewy morning I can be blissfully alone while the house sleeps. And it doesn't take much more effort to go a bit further and enjoy an even greater feeling of solitude.

One of the prominent features of East Lothian's flat arable plain is Traprain Law. A volcanic intrusion, its hard basalt rock was not as yielding as the surrounding geology to the scouring of now long-retreated ice sheets, and so it remained. Millions of years later, although a little weathered and with a huge slice quarried from one flank, it still stands proud. The north-facing slopes are gentler; harebells, burnet saxi-frage, wood sage and yarrow hold their own against a thicker growth of grass. Exmoor ponies have been employed to bite back the more dominant growth, allowing a greater diversity of flora to return. On the southern side it is very different. Lichen-encrusted slabs of rock, with thyme and mosses sprouting from their cracks and fissures, rise up from a protective moat of rough grassland, in turn surrounded by a sea of intensively farmed crops.

I decide to take advantage of a windless evening in July to take some traps to the top of this hill. I'm feeling optimistic, or strong, or both, so furnish myself with four traps: three hanging off my milk-maid's pole, one in my hand. Hoisting a rucksack straining at the seams with the required four batteries onto my back instantly eliminates any spring in my step, and manoeuvring my way over the first stile presents a considerable challenge that erodes a big chunk of that initial optimism. A few hundred metres later, just as the upward ascent begins, my shoulders start to sing out in discomfort. I am too stubborn to downgrade to three traps, but I lower my ambitions slightly and decide to leave the first trap just a little further up, on a small ledge of the lower slopes.

Slightly lighter, I trudge ever upward, getting into a rhythm and allowing myself a quietly muttered curse on every fourth step. The second trap is shed halfway up the western flank, among rocks and moss and a lone ragwort. I sit back for a moment to admire the view, looking across to Edinburgh as the cloudy grey sky starts to swell with a faint pinkness. A meadow pipit flitting in the grass beneath me is my only company. My back is sore and sweaty, but I'm enjoying myself.

Under a now much easier load I gather myself for the final ascent. I want to target the rocky areas high up on the southern face; I've only tried trapping here once before and I'm eager to discover more of the moths which fly around the plateau after dark. The Exmoor ponies are gathered on the top. I wish them a good evening and they observe me for a few moments with a mixture of aloofness and mild curiosity before returning to their evening meal of grass. Seeking out some flatter ledges among the rock before it plummets vertiginously downwards, I position the final traps before finding a comfortable spot to simply sit in the gathering dusk. I wonder about the generations of people stretching back in time to millennia BCE who might once have enjoyed the same seat. How different was their view, how different their contemplation?

The following morning, I return at dawn. A swirling mist has appeared overnight, reducing my world to monochrome grey as I stride, empty-loaded, upwards. No matter about the loss of far-reaching views or missing out on sunrise, it is the moths I'm here for, but I hope the mist stayed away long enough during the night to allow the lights on my traps some reach.

The first two traps have a good selection of moths resting on their egg boxes, my favourite a Figure of Eighty (*Tethea ocularis*). About two centimetres long, it has scribed on each slate-grey wing a series of small white circles clearly denoting an '80': a moth that helpfully wears its name on its wings.

Up on the summit the mist swirls spookily, like something out of *Macbeth*. Occasionally the yolk of the sun is discernible through the milky whiteness, before being smothered once more. The ponies are still here, one moment softly silhouetted against the misty backdrop, the next barely discernible fuzzy forms. Without any distant view, my imagination could take me almost anywhere.

With anticipation heightened by the magical atmosphere I arrive at the first of the summit traps. But any optimism is quickly dampened as the view through the Perspex lid looks disappointingly empty. I wasn't expecting hordes of moths on this bleak plateau, but a few would be nice. As I examine each egg tray in turn my disappointment mounts. Nothing. One tray shelters a single fly, but I'm not in the mood for flies this morning. Near the bottom I finally get a moth. A Dark Arches (*Apamea monoglypha*). A common and ubiquitous species; I've already had several in the traps on the lower slopes. Not the sort of treasure I was hoping for but proof that the trap must have been working; I can't even blame faulty electrics for the poor result. As my hopes switch to the final trap, just a short distance away, I'm stopped. In the corner of the final egg tray is a beauty, matching the colour of the grey cardboard in which it nestles. I almost missed it. A moth I have never seen before. I think I know it from an illustration in a book but without the reference at hand I can't be certain. But what else can it be? Northern Rustic (*Standfussiana lucernea*).

It is soft and furry and looks warm and cosy in the morning's damping, with perfect wings coloured in indistinct gradations of grey like the mist around us. I coax it onto a grey, lichen-encrusted rock where it blends in seamlessly. It isn't colourful, it isn't showing much charisma, but I love it. I perch on the damp rock next to it. This is the first time this species of moth has been seen in East Lothian, though I suspect they have been living here far longer than I have, perhaps noticed by previous visitors to this rocky hill, but never documented.

The mist is gradually lifting, the sun starting to win the battle against water droplets as it always, eventually, does. I coax the Northern Rustic into a more discreet nook, where it will hopefully stay safe from chancing predators, and pack up to move on to the final trap. My early morning finishes with a handful more moths, but none as special as the Northern Rustic.

I've been back to this hilltop with my traps every July since, enjoyed sunsets and sunrises, talked to the ponies, occasionally met other people, but have yet to encounter Northern Rustic again. I hope they are still there. I will keep returning. The excuse: to find another Northern Rustic. The reason: because to be outside somewhere, when the day becomes night and the night becomes day, is a time to cherish. But always, in the recesses of my mind, there is the hope of sharing it with some fine moths.

Acknowledgements

First, thank you to the many amazing people who provide me with entomological inspiration, encouragement and optimism. The natural history community is a wonderful thing to be part of.

Those mentioned here have contributed to the production of this book. Any factual errors are mine. With apologies to anyone I've missed, my thanks go to:

Ashleigh Whiffen who was a key influence in getting this book started. Her unwavering support of my moth investigations at the National Museums of Scotland and encouragement in every entomological endeavour I undertake has been incredible. I apologise that there is less about the wonder and importance of museum entomology collections in this book than there should be.

Justine Patton and Jennifer Vickers were so helpful in providing me with material when I was starting to write. In the paring-down process, your stories fell onto the cutting-room floor, but your early input certainly helped shape the book. I hope one day we will meet in person.

A big thank you to Mark Cubitt, who first encouraged to explore the wilds of East Lothian for moths and has since become a good friend. I hope we have many more single-minded mothy adventures together. Thanks also to Roy Leverton for sharing his vast knowledge and provoking discussions via email; Nigel Voaden for help with micros, enjoyable outings in search of leaf mines and some gentle moth competition, that I will never equal, from the other side of the Forth; and Mark Young, whose unconmoth-ing support and encouragement has been more valuable than he probably realises.

Thanks to other friends and colleagues who have provided stories and information: Tom August, Douglas Boyes, Joe Burman, Nyree Fearnely, David Fearnley, John Harrison, David Hodgson, Marc Holderied, Ellie Lawson, Paul Millard, Don Opitz, Bethia Pearson, David Pickett, Michael Pocock, Zoe Randle, Helen Rowe, Juliette Rubin, Scott Shanks, Mark Shaw, Améline Traschler, Paul Waring, Dave Wild.

Special mention must be given to Douglas Boyes, who tragically passed away during the writing of this book. His research contributed much to our understanding of moths and his generosity and enthusiasm has left a lasting influence on me and so many others. Thank you to his mother, Claire, for reading over his contribution to Chapter 2.

One person appears in this book as an unnamed friend. Donald Smith is my regular companion on entomological trips in East Lothian. I search for moths; he hunts for flies. He has listened patiently to the ups and downs of my book writing journey and regularly offered much-needed encouragement. Thank you, Donald.

ACKNOWLEDGEMENTS

I'm indebted to Clare Conrad from Janklow & Nesbit for turning some half-baked ideas into reality and more. To Louise Haines and Mia Colleran at 4th Estate for their editorial skill and patience. To Sally Partington who worked efficient wonders at the copy-editing stage and Jo Thomson for creating the wonderful illustrations.

Finally, family. Thanks to my parents Maggie and Bruce Ponder for helpful critique on an early draft and some all-important parental endorsement; to my husband Mark and children Tamsin, Gemma and Sam for laughing at me and letting me get on with it. One day, I hope the moth knowledge you've inadvertently absorbed will come in useful.

Map

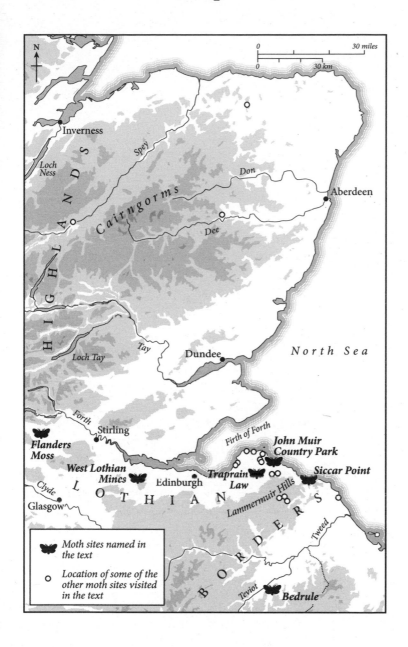

How to Find Moths

Moths can be found almost anywhere but there are techniques that will increase your chances of seeing them. More information is available from the resources listed in the following pages, but some suggestions are outlined here. My best advice is: give it a go!

Light: The simplest approach is to leave an outside or porch light on after dark and wait for moths to come. Either catch them carefully into a pot as they arrive, or check the nearby walls and fences first thing in the morning to see what is resting there. Moths will also come to lighted windows. A white sheet hanging up with a bright torch shining on it can also be used. Once your moth appetite has been whetted, specialist light traps with bright UV-rich bulbs are an excellent way to attract large numbers of moths overnight. These can be purchased from entomological suppliers or, more cheaply, made at home. A number of straightforward designs with instructions are available on the internet.

Sugaring: Heat about 500 ml of brown ale (or cola will do) in a large pan and simmer for five minutes. Stir in about a kilogram of dark brown sugar, followed by a tin of black treacle. Simmer the mixture for a few minutes and then allow it to cool before transferring it to a suitable lidded container for transporting outside. Just before dusk, use a paintbrush to spread small amounts of the mixture onto tree trunks or fence posts at about eye level. As darkness falls, moths will be attracted to the sugar to fuel up for the night ahead. Use a torch to check the sugary patches for moths during the first couple of hours after dusk.

An alternative to painting a sugary mix onto posts is wine ropes. These are made by soaking lengths of rope or scraps of sheet with a syrupy mixture made by heating half a bottle of red wine with a bag of sugar. The sticky rags are draped over branches or hung from fences and checked with a torch just after dusk. At the end of the evening, retrieve the rags and save them to re-soak for another night.

Mashed-up overripe fruit can also be effective at attracting moths, particularly the *Catocala* underwing moths in early autumn. Either mix with your sugaring mixture and paint onto posts, or serve on a plate in the garden.

Flowers: Many flowers that attract butterflies and bees with their nectar by day are also likely to attract moths, both during the day and at night. Look for nocturnal moths guzzling on nectar by shining a torch on flowers in the first couple of hours after dusk. Flowers that emit fragrance after dark are particularly alluring. If you have space, consider growing suitable plants in your garden or in containers on a balcony or porch. Many wildlife organisations provide gardening suggestions and guidance on their websites. As well as providing nectar for adult moths, don't forget to grow some tasty leaves for the caterpillars; they won't eat them all.

Netting: Not all moths are attracted to light, sugar or nectar. Moths flying in the daytime or at dusk can be swept into a net so they can be examined close up for identification, before carefully releasing them. Suitable nets can be purchased from entomological suppliers or, if skills allow, home-made by sewing a large bag from a lightweight fabric. Attach it to a frame made from sturdy wire or, better still, repurpose an unused tennis racquet.

Further information

There are many excellent organisations, journals and websites that provide more information about the natural history of moths and how to record them. Below is just a selection.

Butterfly Conservation: www.butterfly-conservation.org

National Moth Recording Scheme: https://butterfly-conservation.org/our-work/recording-and-monitoring/national-moth-recording-scheme

Atropos magazine: www.atropos.info

The Entomologist's Record and Journal of Variation: www.entrecord.com

UK moths website: www.ukmoths.org.uk

Lepiforum: www.lepiforum.de

Moths of Ireland: www.mothsireland.com

The Lepidopterists' Society: www.lepsoc.org

Lepidopterists' Society of Africa: www.lepsocafrica.org

Societas Europaea Lepidopterologica: http://www.soceurlep.eu

Selected Reading

Allan, P. B. M., *A Moth Hunter's Gossip* (Philip Allan & Co. Ltd., 1937)

Gandy, Matthew, *Moth* (Reaktion Books, 2016)

Hancock, Brian, *Pug Moths of North-west England* (Lancashire and Cheshire Fauna Society, 2018)

Henwood, Barry and Sterling, Phil, illustrated by Richard Lewington, *Field Guide to the Caterpillars of Great Britain and Ireland* (Bloomsbury Wildlife Guides, 2020)

Lees, David C. and Zilli, Alberto, *Moths: Their Biology, Diversity and Evolution* (Natural History Museum, 2019)

Leverton, Roy, *Enjoying Moths* (Poyser Natural History, 2002)

Lowen, James, *Much Ado about Mothing: A year intoxicated by Britain's rare and remarkable moths* (Bloomsbury Wildlife, 2021)

Manley, Chris, *British and Irish Moths: A Photographic Guide*, 3rd Edition (Bloomsbury Wildlife, 2021)

Marjerus, Michael, *Moths*, New Naturalist Library 90 (Collins, 2002)

Marren, Peter, *Emperors, Admirals and Chimney Sweepers: The naming of butterflies and moths* (Little Toller, 2019)

Newland, David, Still, Robert and Swash, Andy, *Britain's Day-flying Moths*, 2nd Edition (WILDGuides, Princeton University Press, 2019)

Randle, Z., Evans-Hill, L. J., Parsons, M. S., Tyner, A., Bourn, N. A. D., Davis, T., Dennis, E. B., O'Donnell, M., Prescott, T., Tordoff, G. M. and Fox, R., *Atlas of Britain and Ireland's Larger Moths* (Pisces Publications, 2019)

Salmon, Michael A., *The Aurelian Legacy: A history of British butterflies and their collectors* (Harley Books, 2000)

Smart, Ben, *Micro-moth Field Tips 1* (Lancashire and Cheshire Fauna Society, Rishton, 2020)

Smart, Ben, *Micro-moth Field Tips 2* (Lancashire and Cheshire Fauna Society, Rishton, 2022)

Sourakov, Andrei and Chad, Rachel Warren, *The Lives of Moths. A Natural History of our Planet's Moth Life* (Princeton University Press, 2022)

Sterling, Phil and Parsons, Mark, *Field Guide to the Micromoths of Great Britain and Ireland* (Bloomsbury Wildlife Guides, 2018)

Waring, Paul and Townsend, Martin, *Field Guide to the Moths of Great Britain and Ireland*, 3rd Edition (Bloomsbury Wildlife Guides, 2017)

Young, Mark, *The Natural History of Moths* (Poyser Natural History, 1996)

Notes

Chapter 1: Getting to Know Moths

1 *Number of species of moth*: Calculating the number of known species of moth is difficult. New ones are discovered each year from regions all over the world, but it can take several years from the initial discovery to the moth being formally described and added to the world list. Meanwhile, species are also going extinct. Further complications arise from 'lumping and splitting'. More detailed taxonomic studies that include genetic sequencing might reveal that what were once considered several different species are in fact variants of the same species; or a number of moths initially considered a single species might turn out to be two or more different species with a superficially similar appearance. An examination of published resources in 2011 cites 157,424 described species: E. J. van Nieukerken *et al.* (2011). Order Lepidoptera Linnaeus, 1758 in *Animal Biodiversity: An Outline of Higher-Level Classification and Survey of Taxonomic Richness*, ed. Z-Q. Zhang. Magnolia Press, Auckland, New Zealand, vol. 3148, pp. 212–221.

2 *Fossil scale discovery by Dutch geologists*: Dutch geologists looking for evidence of pollen by sifting through ancient pond sediments discovered tiny moth scales which they dated to 200 mya. Van Eldijk, T. J. B., Wappler, T., Strother, P. K. *et al.* (2018). A Triassic-Jurassic window into the evolution of Lepidoptera, *Science*

Advances, Jan 10;4(1):e1701568. doi: 10.1126/sciadv.1701568. PMID: 29349295; PMCID: PMC5770165.

3   Biological Classification of Lepidoptera: All living things are organised into a hierarchical series of groups and sub-groups on the basis of their similarities. Whereas once this was based on looking at morphological characteristics, now genetic analysis is also used. At the top of the hierarchy are five Kingdoms: Animals, Plants, Fungi, Protists, Prokaryotes. Each Kingdom is split into different Phyla which in turn are split into different Classes and then Orders. Butterflies and Moths are part of the Animal Kingdom in the Phylum Arthropoda, Class Insecta, Order Lepidoptera. The Lepidoptera are then divided into different Superfamilies and Families. Families are subdivided into different Genera (single: Genus), each of which contains a number of different species. A species can be defined as a group of similar organisms that are capable of breeding with one another to produce fertile offspring. Each species has a unique scientific, or binomial, name which is used throughout the world. This comprises two words: the first indicates the genus to which it belongs and the second is the specific species name. Many moths also have common, or vernacular, names that are widely used, but these are not universal and often vary from country to country.

4   Impact of woodland caterpillars on songbirds: The longest-running study linking breeding success of birds with food supply started in 1947 monitoring great tits and blue tits in Wytham Woods near Oxford. Subsequent work in the same woodland from the Hope Department of Entomology looked at availability of insect prey for birds. Van Noordwijk, A. J., McCleery, R. H., and Perrins, C. M. (1995). Selection for the Timing of Great Tit Breeding in Relation to Caterpillar Growth and Temperature. Journal of Animal Ecology 64(4), 451–458. https://doi.org/10.2307/5648. More recent studies in other places with other species show similar dependency on caterpillars for fledgling success.

5   Number of plants pollinated by moths in UK: MacGregor, C. J., Pocock, M. J. O., Fox, R., and Evans, D. M. (2015). Pollination by nocturnal Lepidoptera, and the effects of light pollution: a review. Ecological Entomology 40(3), 187–198, https://doi.org/10.1111/een.12174

Chapter 2: Like a Moth to a Flame

1 Moth-ing the streetlamps: In the last decade of the nineteenth century, publications such as The Entomological Record and Journal of Variation have numerous articles describing discoveries made at the gas lamps. Arthur Naish, in 1856, describes the competitive sport of catching moths from gas street lamps in Bristol in The Entomologist's Weekly Intelligencer 1:163–164.

2 Moth trap designs: Light traps for moths work by attracting the insects with a light and funnelling them into a box where they find it hard to escape. The most effective bulbs emit a high proportion of UV radiation. Mercury-vapour and Actinic bulbs are two of the most popular, but although still widely available these are now no longer manufactured. Instead, improving LED technology looks set to slowly replace them. Many entomologists believe the size of moth captures will reduce as a result. Trap design can vary. The Robinson trap is generally considered to be most effective, but other popular designs are the Skinner trap – a box with two inward-sloping vanes to channel the moths into the trap – and the Heath trap; a much smaller tub and funnel-type design. Both Skinner and Heath traps suit situations where portability is important.

3 A review of many years of mark-recapture work on moths: Waring, P. (2016). Mark and recapture work on moths. Atropos 57:22–33.

4 Light pollution has become a global problem: 83 per cent of the world's population is affected by light pollution (or 99 per cent of North America and Europe's population) and 23 per cent of the Earth's land surface suffers from light pollution. As well as increasingly reported negative impacts on a wide range of biodiversity this also affects human health and well-being. Falchi, F., Cinzano, P., Duriscoe, D. et al. (2016). The new world atlas of artificial night sky brightness. Science Advances, 2(6), e1600377. https://doi.org/10.1126/sciadv.1600377

5 Moths that show a strong attraction to light at night are also suffering greatest declines: van Langevelde, F., Braamburg-Annegarn, M., Huigens, M.E. et al. (2018). Declines in moth populations stress

the need for conserving dark nights. *Global Change Biology.* 24: 925–932. https://doi.org/10.1111/gcb.14008

6   *Douglas Boyes' work on how light pollution could be impacting moths:* Boyes, D. H., Evans, D. M., Fox, R. *et al* (2021). Is light pollution driving moth population declines? A review of causal mechanisms across the life cycle. Insect Conservation and Diversity, 14: 167–187. https://doi.org/10.1111/icad.12447

7   Boyes, D. H., Evans, D. M., Fox, R. *et al.* (2021). 'Street lighting has detrimental impacts on local insect populations'. *Science Advances,* 7(35), p.eabi8322. https://doi.org/10.1126/sciadv.abi8322

Other selected publications by Douglas Boyes:

Boyes, D. H. and Lewis, O. T. (2019). Ecology of Lepidoptera associated with bird nests in mid-Wales, UK. *Ecological Entomology,* 44: 1–10. doi:10.1111/een.12669, and Boyes, D. H. (2018). Lepidopteran lodgers: recording moths from bird nests. Atropos, 62: 42–50.

Boyes, D. H., Fox, R., Shortall, C. R. and Whittaker, R. J. (2019). Bucking the trend: the diversity of Anthropocene 'winners' among British moths. *Frontiers of Biogeography,* 11: e43862. doi:10.21425/F5FBG43862

Chapter 3: In Broad Daylight

1   *The entomological legacy of Miriam Rothschild:* Van Emden, H. and Gurdon, Sir John (2006). *Biographical Memoirs of Fellows of the Royal Society.* 52, 315–330, https://royalsocietypublishing.org/doi/pdf/10.1098/rsbm.2006.0022

Chapter 4: Counting Moths

1   *Citizen science:* The *Oxford English Dictionary* defined citizen science in 1989, but the concept covers a range of purposes and approaches and finding an all-encompassing definition is a challenge. For more see:

Haklay, M., Dörler, D., Heigl, F., *et al.* (2021). 'What Is Citizen Science? The Challenges of Definition' In: Vohland K. *et al.* (eds)

The Science of Citizen Science. Springer Nature, 2021. https://doi. org/10.1007/978-3-030-58278-4_2

2 Randle, Z., Evans-Hill, L. J., Parsons, M. S. *et al.* (2019). *Atlas of Britain and Ireland's Larger Moths.* Pisces Publications.

3 The Rothamsted Insect Survey has been running since 1964. Its data include larger moths, and represent the most comprehensive standardised long-term data on insects in the world. https://www. rothamsted.ac.uk/insect-survey

4 Fox, R., Dennis, E. B., Harrower, C. A. *et al.* (2021). *The State of Britain's Larger Moths 2021.* Butterfly Conservation, Rothamsted Research and UK Centre for Ecology & Hydrology, Wareham, Dorset, UK.

5 *Former abundance of Tiger Moth caterpillars:* Newman, E. (1871). *An Illustrated Natural History of British Butterflies and Moths.* London: W. H. Allen. The moth guides published over a hundred years ago are a rich source of information on the natural history of our moths. They contain information that is often left out of modern-day guides in the interests of space saving and also offer a glimpse of the abundance and distribution of moths in the past.

Chapter 5: Caterpillar Hunt

1 *Macro-moths and micro-moths:* Macros is a generic description for the larger moth species. Our smaller moths are often referred to as micros i.e. everything that isn't a macro-moth. The original split was made on evolutionary age, with the more primitive families (also generally smaller-sized moths) classed as micro-moths. This means that groups such as Swift Moths, Burnets and Clearwings fall within the micro-moths grouping despite containing some of our largest species. Modern usage is less rigid about evolutionary lineage and tends to include these larger micro-moths in with the macros.

Chapter 6: The Change

1 *Maria Sibylla Merian:* For an excellent account of Maria Sibylla Merian's life and the scientific activity surrounding the unravelling

of the process of metamorphosis, see: Todd, K. (2007). *Chrysalis: Maria Sibylla Merian and the Secrets of Metamorphosis.* I. B. Tauris.

2 *An account of how to hunt for moth pupae:* Greene, Joseph (1870). The insect hunter's companion, being instructions for collecting and preserving butterflies and moths and comprising an essay on pupa digging. 2nd edition edited by Edward Newman. London, J. Van Voorst. https://www.biodiversitylibrary.org/item/255995#page/45/mode/1up

Chapter 7: Amorous Aromas

1 *Distance over which sex pheromones act in the field:* This is hard to quantify as it is difficult to prove if a moth is travelling purposefully towards a scent it has detected. Most field studies have looked at crop pest species, and have shown the attraction can work over hundreds of metres and is most effective from downwind of the pheromone source. Laboratory studies show that a male's antennae can respond to minute concentrations of pheromone chemical in the air, and we know some moth species have the capability to range over hundreds of kilometres. Even so, reports of male silk moths travelling over 10km to a calling female are remarkable.

2 *The Life of the Caterpillar* by John Henri Fabre has been translated into English by Alexander Teixeira de Mattos: http://www.efabre.net/chapter-xi-the-great-peacock

Jean Henri Fabre (1823–1915) was a popular teacher and scientist, best remembered for his study of entomology. He wrote several books which convey a joy of nature and demonstrate well the scope of simple observation and an inquisitive mind for making new discoveries.

3 *Studies to show sensible use of pheromone lures doesn't have a negative impact on moths' lives:*

Oleander, A., Thackery, D. and Burman. (2015). The effect of exposure to synthetic pheromone lures on male *Zygaena filipendulae* mating behaviour: implications for monitoring species of conservation interest, *Journal of Insect Conservation* 19, 539–546. https://doi.org/10.1007/s10841-015-9775-4

Thackery, D. and Burman, J. (2016). The effects of synthetic pheromone exposure on female oviposition and male longevity in Zygaena filipendulae (Linnaeus, 1758) (Lepidoptera: Zygaenidae, Zygaeninae). *Entomologist's Gazette*, 67.

4 *Responsible use of pheromone lures:* This includes only leaving them out for short periods of time and avoiding repeated use at sites where the moths occur. Butterfly Conservation has produced a Guidance Note on the use of Pheromone Lures for Recording Moths is available here: https://www.angleps.com/A_Brief_Guidance_Note_on_the_use_of_Pheromone_Lures_for_Recording_Moths.pdf

Chapter 8: Sweet Tastes

1 *Darwin and Wallace's prediction of the coevolution of a long-spurred orchid and long-tongued moth:* For more details including the more recent literature that proved these men's predictions were indeed correct, see: Arditti, J., Elliott, J., Kitching, I. J. and Wasserthal, L. Y. (2012). 'Good Heavens what insect can suck it' – Charles Darwin, Angraecum sesquipedale and Xanthopan morganii praedicta, *Botanical Journal of the Linnean Society*, Volume 169, Issue 3, July 2012, pp. 403–432, https://doi.org/10.1111/j.1095-8339.2012.01250.x
Wallace's Sphinx moth *Xanthopan praedicta* is endemic to Madagascar. In 2021 DNA analysis revealed that it is a distinct species from *Xanthopan morganii* which occurs on the African mainland, rather than a sub-species *Xanthopan morganii praedicta* as previously thought.

2 *Are moths in the tropics more colourful than those of temperate regions?:* Dalrymple, R. L., Kemp, D. J., Flores-Moreno, H. et al. (2015). Birds, butterflies and flowers are not more colourful in the tropics than at higher latitudes. *Global Ecology and Biogeography*, 24: 1424–1432. https://doi.org/10.1111/geb.12368

Chapter 9: Evasive Action

1 *Research on the interactions between bats and moths:* In North America, two research groups in particular, one headed by Prof Akito Kawahara

(University of Florida) and the other by Prof Jesse Barber (Boise State University), are exploring these interactions in different ways. As well as being fascinating in its own right, such research helps explain why some moths look or behave the way they do and how their different adaptations might have evolved over time. This can also help us understand how man-made changes to the natural world might impact moths and the other animals that depend on them.

2 *Gleaning behaviour in bats:* A number of echolocating bat species can apparently detect entirely motionless prey. By converting ultrasound echoes into visual representations, researchers have shown how bats might detect 'echo shadows' made by moths resting on flat surfaces. These shadows are much more obvious on smooth rather than rough textured substrates. See: Clare, E. L. and Holderied, W. (2015). Acoustic shadows help gleaning bats find prey, but may be defeated by prey acoustic camouflage on rough surfaces. *eLife* 4:e07404 https://doi.org/10.7554/eLife.07404

3 *Using genomics to understand when different traits have arisen during the evolution of Lepidoptera:* Kawahara, A. Y., Plotkin, D., Espeland, M. et al. (2019). Phylogenomics reveals the evolutionary timing and pattern of butterflies and moths. *Proceedings of the National Academy of Sciences.* USA, 116 (45) 22657-22663

4 *Replicating the structure of moths' wings in building materials:* In the UK a team of scientists at the University of Bristol, led by Prof Marc Holderied, is investigating the biophysical properties of Lepidopteran scales and how their structure helps them evade detection by bats' ultrasonic hunting strategies. One exciting outcome of understanding how moth wings absorb sound energy is its potential for the development of ultra-thin sound-absorbing panels for use in buildings or vehicles. For example, having very lightweight sound-absorbing panels in planes, cars and trains would lead to lighter vehicles, reduced fuel consumption and thus lower CO_2 emissions. For more details, see Neil, T. R., Shen, Z., Robert, D. et al. (2022). Moth wings as sound absorber metasurface. *Proceedings of the Royal Society.* A478:20220046. https://doi.org/10.1098/rspa.2022.0046 and references therein.

5   *Deflecting bat attacks with wing tails*: More details can be read in the
following publications, and the references therein provide further
insight into this field of study:
Barber, J. R., Leavell, B. C., Keener, A. L. *et al.* (2015). Moth tails
divert bat attack: Evolution of acoustic deflection. *Proceedings of the
National Academy of Sciences*, 112(9), 2812–2816. https://doi.org/10.1073/
pnas.1421926112
Rubin, J. J., Hamilton, C. A., McClure, C. J. W. *et al.* (2018). The
evolution of anti-bat sensory illusions in moths. *Science Advances*.
July 4; 4(7):eaar7428. doi:10.1126/sciadv.aar7428. PMID: 29978042;
PMCID: PMC6031379.
6   *Parasitoid moths*: Two species of Pyralid moths in the genus Chalcoela
from North America are parasitoids of Paper Wasps *Polistes sp*. Their
larvae feed on the larvae and pupa of the wasps. The caterpillars
spin a characteristic web in the cells of the wasp nest.

Chapter 10: Masquerade

1   *Kettlewell's experiments on Peppered Moths*: For a good review of
Kettlewell's work, the ensuing controversy and how it has been
addressed subsequently see:
Majerus, M. E. N. (2009). 'Industrial melanism in the peppered
moth, Biston betularia: an excellent teaching example of Darwinian
evolution in action', *Evolution: Education and Outreach* 2, 63–74. https://
doi.org/10.1007/s12052-008-0107-y
Majerus, M. E. N. (2009). Non-morph specific predation on
peppered moths (*Biston betularia*) by bats. *Ecological Entomology*.
33, 679–683. https://doi.org/10.1111/j.1365-2311.2008.00987.x and
references therein.
2   *The Natural History Museum's online data portal*: Explore and download
the museum's research and collections data. NHM data portal home
page: https://data.nhm.ac.uk See also Scott, B., Woodburn, M.,
Vincent, S. *et al.* (2019). The Natural History Museum Data Portal,
in *Database, the Journal of Biological Databases and Curation*, https://doi.
org/10.1093/database/baz038

Chapter 11: Long-haul Travel

1 *Dispersal capabilities of resident moths*: Some of our resident moths are more dispersive than others. For example, those with caterpillars that feed on ephemeral plants must be able to seek out areas of foodplant away from their natal patch, whereas those that rely on long-lived trees can be more sedentary. With climate change and its impacts now happening so rapidly, moths that are able to disperse easily to seek out new habitat are likely to do better than more sedentary species.

2 *Moth migration*: Studies with Silver Y show they emerge from their pupa with the 'urge' to travel. These young adults take to the air at dusk, but only if the wind direction and conditions are suitable do they commit to the long journey; otherwise they drop to the ground and wait for better conditions. Over the coming days hormones are released and they reach sexual maturity, after which the urge to travel switches to an urge to reproduce. This means they only have a limited window of time to migrate. The distance they can travel is influenced both by weather conditions and the length of this pre-reproductive period.

Perhaps the most impressive long-distance moth migrator is the Bogong moth (*Agrotis infusa*) of Australia. Unusually for a migrating insect, rather than sharing the back-and-forth migration over successive generations, each moth makes both long-distance journeys. Newly emerged moths travel to escape the dry summer heat in the vast lowland plains where they breed, spending these hotter months in the relative cool of mountain caves. As summer ends, they leave the caves and make the return journey to the lowland plains they originated from, where they breed. Fascinating research on Bogongs in the laboratory and field is shedding light on the ecology, physiology and genetics of their migratory behaviours. In addition to helping unravel the mechanism of migration, this research is contributing important information relevant to conservation efforts in the face of environmental change.

3 *Predicting and sharing migrant moth activity:* In the UK many moth recorders share information about migrant moth finds. The Twitter account @MigrantMothUK shares weather charts and predictions. Atropos has a 'Flight Arrivals' page where sightings can be posted and viewed at http://www.atropos.info/flightarrivals

These weather charts can also be used to 'back-track' the paths of migrants once they have arrived on our shores, providing information on their likely location of origin.

4 *The Diamond-back moth forecasting system from the Warwick Crop Centre at the University of Warwick:* https://warwick.ac.uk/fac/sci/lifesci/wcc/research/pests/plutella

Chapter 12: Coping with Cold

1 *Wingless moths:* Around the world a small number of moth species have reduced wings and cannot fly. In most cases it is only the females that are flightless. After mating, they go on to lay their eggs nearby. As well as species that are adults in cold climates, flightless females are often found in species that live in isolated habitats such as small oceanic islands. Here flying brings a risk of being blown into unsuitable habitat for egg-laying or for the caterpillars to develop.

Chapter 13: Then and Now

1 *The value of carefully curated insect collections:* Colvin, M. (2014). Entomological Collections – Their Historic Importance and Relevance in the 21st Century [Online]. Available from http://www.dispar.org/reference.php?id=92 [Accessed June 19, 2022].

2 *Using automatic recognition technology for identifying moths:* This is a rapidly advancing field of research offering enormous potential for biodiversity monitoring. The following publications offer an introduction to some areas of study:

Terry, J., Roy, H. and August, T. (2019). Thinking like a naturalist: Enhancing computer vision of citizen science images by harnessing contextual data. *Methods in Ecology and Evolution* 11. https://doi. org/10.1111/2041-210X.13335.

Bjerge, K., Nielsen, J. B., Sepstrup, M. V. *et al.* (2021). An Automated Light Trap to Monitor Moths (Lepidoptera) Using Computer Vision-Based Tracking and Deep Learning, *Sensors* 21, no. 2: 343. https://doi. org/10.3390/s21020343

Moth and Caterpillar Index

General Index